Ghosts in the Machine

Ghosts in the Machine
Rethinking Learning Work and Culture in Air Traffic Control

Christine Owen

CRC Press
Taylor & Francis Group
Boca Raton London New York

CRC Press is an imprint of the
Taylor & Francis Group, an **informa** business

CRC Press
Taylor & Francis Group
6000 Broken Sound Parkway NW, Suite 300
Boca Raton, FL 33487-2742

© 2018 by Taylor & Francis Group, LLC
CRC Press is an imprint of Taylor & Francis Group, an Informa business

No claim to original U.S. Government works

Printed on acid-free paper

International Standard Book Number-13: 978-1-4094-5290-4 (Hardback)

Library of Congress Cataloging-in-Publication Data

Names: Owen, Christine, 1958 April 18- author.
Title: Ghosts in the machine : rethinking learning work and culture in air
traffic control / Christine Owen.
Description: Boca Raton : CRC Press, Taylor & Francis Group, 2017.
Identifiers: LCCN 2017014537| ISBN 9781409452904 (hardback : acid-free paper)
| ISBN 9781315584706 (ebook)
Subjects: LCSH: Air traffic controllers--In-service training. |
Organizational behavior. | Air traffic control--Vocational guidance.
Classification: LCC TL725.3.T7 O94 2017 | DDC 387.7/40426--dc23
LC record available at https://lccn.loc.gov/2017014537

Visit the Taylor & Francis Web site at
http://www.taylorandfrancis.com

and the CRC Press Web site at
http://www.crcpress.com

Contents

List of figures

List of tables

About the author

Christine Owen is a researcher with a focus on organisational behaviour and learning. Christine has an established and growing reputation as a human factors researcher and facilitator within emergency management. Her research investigates communication, co-ordination and teamwork practices in high-technology, high-intensity and safety-critical work environments.

chapter one

Introduction

Some two decades ago, I was working in a professional development context, providing assistance to air traffic control instructors who were working at an Air Traffic Control training college. I had been doing so for a couple of years and enjoyed the enthusiasm the controller-instructors displayed for their role in facilitating the learning of others. Despite this, however, I also noticed that nothing much seemed to change in the on-the-job training that occurred at the console. It seemed that somewhere between the training room and the simulation console – as well as the consoles at the air traffic control centres around the country – the insights and commitment gained to working differently have evaporated.

Then, one of the training programs I was facilitating was worthy of note for the complete and utter failure on my part to convince a group of controller-instructors of the value of applying adult learning principles to their instruction. In short, it did not matter how much I extolled the virtues of adult learning principles – these controller-instructors were having none of it! Tired and frustrated on the final day, I asked why they were not interested.

They explained to me that I had fundamentally missed the point of air traffic control instruction. First, air traffic control was not something that could be learned: *'You've either got it or you haven't'*. What I was attempting to do was to *'teach the unteachable'*. From their perspective, air traffic control was not something that could be taught. Air traffic controllers were, in their words, *'born and not made'*. Therefore, none of the instructional strategies discussed on the course were of value. Moreover, using them would be dangerous! This was because engaging in instructional practices that enhanced learning and made the job of learning air traffic control easier could result in weak trainees getting through the system. For this instructor cohort at least, training was akin to a rite of passage, as trainees had to 'prove' their worth by getting through the course – in some circumstances – almost despite the instructor. Their job was not to facilitate learning. It was instead to weed out weak trainees and to be a gatekeeper for the profession.

As a frequent flyer, these opinions left me deeply conflicted. On the one hand, I want any aircraft I am travelling on as a passenger to be controlled by someone who is fully competent in both routine and non-routine problems. On the other, I want them to have the courage and strength of will to be able to manage challenges as they arise – the same for the pilots! Yet all expertise requires learning. Studies on expertise acknowledge the role of ability but are quick to point out that while it is a necessary compo-nent, it is insufficient for expertise to develop. What is needed is practice and motivation. Also important is insight and reflection on challenges for improvement and development.

Moreover, what were the risks of these viewpoints? Given that not all instructor-controllers I had met or worked with held the same views as this cohort, I was also interested in why others held different views and how both of these different perspectives came to be – in fact – learned norms of behaviour, complete with expectations and consequences for behaviour.

And so began my journey into trying to understand the role of organ-isational cultures in air traffic control work and how these cultures devel-oped. For an outline of the research conducted and reported in this book, please see Appendix A.

Why ghosts?

Gilbert Ryle was one of the first authors to coin the term 'Ghosts in the machine' to characterise the problem with the concept of mind and how we understand the world. Unlike the Cartesian dualist account of the mind as separate from the body, the argument here is that our thinking does not occur independently but is connected both to our bodies, the interactions we have with others and to our environment, which – in this case – includes the technologies with which we interact. Drawing further on this idea, my claim is that our ways of thinking in workplace contexts are influenced by our social histories of organising around these technol-ogies-in-use and other artefacts.

The title *'Ghosts in the machine'* is used here to draw attention to how organisations comprise people who in turn shape – and are shaped by – their ways of organising. These include organisational structures, such as how tasks are divided between labour, and rules of accountability and authority as well as the applications of technologies-in-use. Job roles and tasks concentrate individual and collective experiences in particular ways. This is because these features concentrate workplace experiences and enculturate people into established ways of acting. Over time people collectively develop routines which then become habits and work norms, establishing collective meaning and identity.

This argument is in keeping with Leontiév (1978), who illustrated how actions are afforded by socially created artefacts. As artefacts change, so too do the work practices they embody. In this way, workplace cultures are created through the relationships between people and those artefacts as they co-produce their activities.

In sum, opportunities for the development of specific work cultures are shaped by the contextual features of work organisation that have also been socially constructed.

My claim in this book is that as processes of technology-mediated organizing develop and change, so too do the cultures that have previously formed. However, these cultures have a transitional 'half-life' that enables both coherence as well as the emergence of contestation between cultures in the organisation.

High-3 work and aviation

While the book is focussed on one occupational group – that of air traffic control (ATC) – the insights should be of interest to scholars interested in a range of other industries that share similar characteristics. In particular, the insights from these observations should be helpful to anyone working in other high-technology, high-intensity and high-reliability (High-3) industries. Work found in High-3 organisations are observable in virtually all industry sectors, although the more-developed examples are to be found within transportation (e.g. air and sea), the military, police, emergency services (e.g. ambulance and firefighting), health services (e.g. operating rooms), manufacturing (e.g. chemical industries) and key elements of the finance sector (e.g. the stock exchange). Moreover, the proportion of organisations characterised by these forms of work is growing as part of the growth of a knowledge-based economy, as is our reliance on them (Powell and Snellman 2004).

As an example of a safety-critical work environment, the aviation industry has advanced many ideas and models to assist in understanding human performance in complex systems.

The aviation industry was born from the risk-taking of individuals attempting to do something previously not possible. The stereotypical images of success within the aviation industry, characterised by the aircraft pilot, 'a single, stalwart individual, white scarf trailing, braving the elements in an open cockpit' (Helmreich and Foushee 1993, p. 4), have led to acceptance – indeed celebration – of norms associated with 'independence, machismo, bravery, and calmness under stress' (Helmreich and Foushee 1993, p. 4). The history of these experiences led to assumptions that successful performance depended on individual activity and not team effort, even when it can be determined that success was based on teamwork (Malakis et al. 2010). In the aviation industry, this view began to change over time, when investigations into jet aircraft accidents began drawing conclusions that 'pilot error' was more likely to reflect failures in team communication and coordination than deficiencies in an individual's technical skills (Murphy 1980). The awareness of the importance of teamwork and interpersonal communication has continued. In investigating incidents and accidents in the United States, it was found that that the most common reasons for communication failures were not because of the unavailability of needed information, but rather because (1) the person who had the information did not think it necessary to transfer it or (2) the information was transferred incorrectly (Hartel and Hartel 1995).

A further change occurred also when more attention was given to the role of context in mistakes made in High-3 work environments. The role of organisational elements in High-3 work was given prominence in 1987, in the maritime world, with Mr Justice Sheen's assessment of the causes of the capsize of the 'Herald of Free Enterprise'. After acknowledging the errors of the crew, he went on to say that the causes of the disaster lay higher up the organisational hierarchy, where a culture of 'sloppiness' appeared evident (Sheen 1987). In the aviation industry, Commissioner Moshansky's (1992) inquiry into the crash at Dryden, Ontario produced a similar judgement and signalled the recognition of the role of organisational structures in the analysis of accidents. When introducing his findings, Commission Moshansky wrote:

> The accident at Dryden... was not the result of one cause but a combination of several related factors. Had the system operated effectively, each of the factors might have been identified and corrected before it took on significance. It will be shown that this accident was the result of a failure in the air transportation system. (Moshansky 1992, in Maurino et al. 1995, p. 2)

Continuous learning and conscious inquiry

In the kind of High-3 work environments represented by ATC, that is, high-intensity, high-technology and high-reliability contexts (e.g. Klein et al. 1995; Vaughan 1997, 2006; Flin 1998; Weick 2012), the practice of conscious inquiry is important in two respects. The first is that in High-3 work environments the consequences of mistakes are so serious that they are unacceptable. Secondly, High-3 work environments are necessarily complex and the practice of conscious inquiry enables individuals to collectively pool the resources they have. These two factors are also connected. Weick (2012) argued that as the complexity of organisations and their technology increases, High-3 work environments become more susceptible to accidents. This is particularly so because the major learning strategies available in other organisations – such as trial and error – are not available in High-3 organisations. One of the reasons for propensity for mistakes in complex organisations is because 'the humans who operate and manage complex systems are themselves not sufficiently complex to sense and anticipate the problems generated by those systems' (Weick 1987, p. 112). Weick concluded that culture can play an important role in enhancing an organization's reliability and its capacity to learn from mistakes.

In this chapter, developing Weick's ideas further, what is required for successful working in environments aiming for high-reliability are work practices of conscious inquiry. Such practices include the ways in which individuals and groups are encouraged to observe, inquire and actively bring their conclusions to the attention of others (Westrum 1997). Conscious inquiry involves practices involved with reflecting, making sense and sharing suggestions with others about enacting new ways of doing.

This has led some experts working within the High-3 field to examine closely the flow of information within organisations (Hollnagel and Woods 2005) and to advocate for the creation of 'generative' organisations, where people can think and communicate effectively.

The skills of conscious inquiry include the practices of informal learning, including being able to ask the right questions, sharing observations, seeking alternative perspectives, assertively challenging a particular opinion, seeking clarification, sharing information and consulting and collaborating. Requisite variety is then enabled because such learning-related behaviours make possible the expansion of the number and variety of experiences, and this increases the number of possible options that might be available.

While changes in thinking about learning and work are reflective of a worldwide phenomenon, their importance to the aviation industry cannot be overstated. In this industry, performance and reliability are generally understood in terms of safety. Safety in aircraft performance has improved dramatically as a result of improvements to the technical and mechanical aspects of both aircraft and air traffic management systems over the past few decades. Although flight is one of the safest modes of transport, and while technical and mechanical faults have declined as a source of errors and accidents, human errors have remained constant over time. This means that between 60% and 80% of aviation accidents can be attributed to human error, most notably interpersonal communication failure (Hartel and Hartel 1995). Hence, enhancing communication, especially between individuals and groups, is particularly important to those involved with this industry.

Rethinking learning and ghosts

The link between ghosts in the machine and rethinking learning is underpinned by socio-cognitive learning theories. According to these theories, the body and the brain, the self and the environment are not dualisms as was once assumed. Learning and experience are integrally linked to situation and context. For example, rather than viewing a person as being 'in' an environment 'just like a cherry in a bowl' (Bredo 1994, p. 28), the activities of person and environment, and of the mind and body, are argued to be part of a mutually constructed whole. 'The inside-outside relationship... which is generally presupposed in a symbol-processing view of learning, is replaced by a part-whole relationship' (Bredo 1994, p. 28). From this perspective, experience involves the self in relationship with the environment, and the mind is not something separate from body or environment but part of a mutually shaping cycle of interaction (brain *and* body, self *and* environment).

Traditional psychological approaches treated the environment or 'context' as something of an 'omnibus category that allowed the analyst to point to factors outside of the psychological task itself as contributors to performance' (Cole et al. 1997, p. 5). However, this view is inadequate because it treats context as a pre-set environment that influences behaviour, rather than as a set of resources people use as they create cognition and culture in ways that are constantly shifting and dynamic (Cole et al. 1997).

In this book, ways of operationalising contexts include the institutional structures that influence everyday action and in the meaning participants give to their interaction. In organisational theory terms, contexts have most commonly been conceptualised in terms of organisational structures and cultures. An organisation's structure is based on a formal system of interlocking roles and relationships between roles and can be seen in forms of division of labour, departments, hierarchy, policies and rules and coordination and control mechanisms (Tosi and Pilati 2011). An organisation's structure is also evident in the external relationships between the organisation and its environment. In addition to the formal systems, every organisation has a set of interrelated informal systems that influence behaviour. These are described as organisational culture or sometimes as communities of practice (Wenger et al. 2002). Organisational culture is a set of understandings shared by a group of people that are largely tacit among members. Values, beliefs, attitudes and norms are used by members to justify certain decisions and behaviour. In practice, both structures and cultures are interwoven. Therefore, learning in the workplace is embedded within structures and cultures of the organisation.

Structure as an element of organisation

An organisation's formal structure refers primarily to the patterns and regularities in its division of labour by task or function, hierarchies of authority and control mechanisms. Organisations also have informal mechanisms that pattern relationships and these are cultures. Organisational cultures are defined as the 'habits, folkways and norms that shape action' (Westrum 1993, p. 401) and the 'set of understandings or meanings shared by a group of people' (Loui 1986, p. 74). In practice, both structures and cultures are interwoven.

Historically, the major eras in management-oriented approaches to work organisation have included 'Taylorism', emphasising the elimination of worker inefficiency. Taylorism, for example, provided early attempts to examine the structure of work organisation by detailing how the modification of rules, divisions of labour and hierarchies can enhance worker productivity. Taylorism led to a focus on work-tasks and structure and hierarchies of wage compensation. Although there is no one best way, most organisational theorists have shifted to a contingency approach (Van de Ven et al. 2013). Contingency theory suggests that the most appropriate set of structures and processes for an organisation will be contingent on finding the best fit with the environment. Organisational problems are characterised as a misfit between the organisation's structure and the environment within which it operates. Three dimensions have been identified as key features of organisational structure (Harper 2015) that enable or constrain fit with the environment.

First, organisations will vary in terms of their complexity, that is, the breadth of activities, job functions and the number of levels within an organisational hierarchy. Differentiation is the process by which people and resources are allocated to different roles, tasks and functions, as evidenced in the organisation's division of labour, and integration refers to the strategies used to coordinate those tasks, roles and functions (Hill et al. 2014). Organisations have long been known to require various forms of 'differentiation' in work activity to achieve to their goals. The foundation of differentiation is the 'job role', first coined by Lawrence and Lorsch (1969). The job role is a set of task-related behaviours required of a person in accordance with their position within the organisation. Simple organisations do not require a high degree of differentiation in the division of labour, whereas complex organisations are likely to require differentiation between many job tasks involving divisions of labour, as people specialise within particular jobs and job roles. Differentiation is organised structurally through mechanisms such as authority and control, that is, the power to make people accountable for their actions, to make decisions about resources and to coordinate work activity. Child (2015) distinguished between differentiation found in an organisational hierarchy (vertical differentiation) and that found in the way in which tasks and roles are organised into subunits, functions or departments (horizontal differentiation). Differentiation can be contrasted with that of integration, which refers to the structures used to coordinate various tasks and functions. One of the earliest studies of integrative mechanisms and their role within organisation was undertaken by Gailbraith (1979). Gailbraith (1979) posited that as organisations increased in complexity, and they needed more and more integrating mechanisms to be effective. The integrating mechanisms examined by Gailbraith (1979) included teams, taskforces, liaison roles and departments with the responsibility of providing an integrative function. Contemporary theorists of organisational design (e.g. Hill et al. 2014) contend that the challenge for organisations is to find ways of balancing differentiation with integration.

Second, organisations will vary in the degree of formalisation used to structure work activities, that is, in the use of policies, procedures and rules that constrain the choices of members in their work. Formalisation is the degree to which use is made of written rules and procedures to achieve standardised operations and conformity. In highly formalised operations, the discretion and autonomy of members is limited. In less formalised organisations, there is more freedom to exercise choice. Organisations relying heavily on formalised procedures are likely to be centralised, whereas organisations that do not are likely to be decentralised. Centralisation refers to the degree to which authority, control and activity are distributed vertically within the organisation (Hall 2002). Authority has been decentralised when key decisions can be made by organisational members at a range of levels within a hierarchy. Activity is centralised when work occurs in a central location, rather than being dispersed through a range of locations. Lawrence and Lorsch (1969) found that when the environment is perceived as unstable and uncertain, organisations are more successful if they are less formalised and more decentralised. However, this will clearly depend on the nature of the work undertaken in the organisation. Organisations characterised by high-reliability, high-technology and high-intensity need high degrees of formalisation, despite operating in uncertain and complex environments (Von Glinow and Mohrman 1990). This is because workers operating within such workplaces need clearly defined protocols or procedures to follow in the event of the unexpected.

Third, organisations will vary in the degree of learning. The notion of collective learning was explored by Nonaka and Takeuchi (1995). They argued that collective learning is particularly important in generating continuous innovation. In their view, the Japanese companies they have studied have a different understanding of knowledge, distinguishing the tacit and the explicit. The tacit includes both technical 'know-how' and beliefs and perspectives, images of reality, while explicit knowledge encompasses the formal and systematic learning, which can be communicated as universal principles or codified procedures. They argued that tacit knowledge is not easily communicated, is learned from direct experience, through both mind and body and involves paying attention to the less formal and systematic side of understanding. Nonaka and Takeuchi (1995) conclude that Japanese companies create new knowledge through 'the conversion of tacit knowledge to explicit knowledge' (p. 11) and thus enhance collective learning within the organisation.

Organisational learning has two features that advance it beyond individual learning: communication and shared interpretation (Weick 2012). Although organizational learning occurs through individuals, it would be a mistake to conclude that organizational learning is nothing but the cumulative result of their members' learning. 'Members come and go and leadership changes, but organizations' memories preserve certain behaviours, mental maps, norms and values over time' (Hedberg 1981, p. 3, cited in Brown et al. 2006). This occurs because of the collective nature of learning; it is not simply based on the sum of the individuals within the organisation. Individuals in organisations, through being socialised into certain values and norms, continue to reproduce patterns of thinking and acting. It is contended that how learning translates from individuals to organisations critically depend on organisational culture (Schein 1996), where shared norms and values indicate organisational rather than individual learning.

One of the key features of organisational learning is the focus on the sharing of information and knowledge across individuals and groups (Gherardi 2009). Whilst individual learning can be shared, it is not necessarily so. However, organisational learning is inherently interactive, interpretive and integrative (Dixon 1999).

Another problem involved in organisational learning is 'unlearning' or forgetting past behaviour that is redundant or unsuccessful (Tsang and Zahra 2008). As situations change, what is regarded as relevant knowledge also changes. The challenge for organisations, then, is to find ways of discarding obsolete and potentially misleading knowledge.

Culture as an element of organisation

Some studies of organisational culture treat it as an homogeneous variable that can be imposed upon the organisation from the top-down (e.g. Wellins et al. 1991). This is the basis for much of the 'corporate culture' literature (Kotter 2008). Other studies (e.g. Loui 1986; Alvesson 2012) point to the heterogeneity of organisational cultures. Such heterogeneity is important because understanding why patterns of difference have arisen would provide evidence of what is valued within the organisation and what is learned in work activity within different groups. Because of this heterogeneity of group membership in organisations, cultures can consist of shared, partly shared, non-shared and/or contested values, beliefs and norms (Gherardi 2009).

It is contended that workplace cultures are revealed in the way people communicate their understanding about their work (Scheeres and Rhodes 2006; Schein 2009; Alvesson 2012); their shared (implicit) norms of behaving (Balthazard et al. 2006); the language they use (Smiricich 1983); the stories they tell (which reveals what they want to remember and what they forget) (Trice and Beyer 1984; Czarniawska 1997) and the stereotypes they employ to account for in-group and out-group group membership (Fine 1996; Holland and Lave 2001; Lawrence 2006). Thus, cultures tie the actions of individuals to a particular group (or groups) and reveal, through justifications for group membership and the informal language that is used, what is collectively valued within the group (and what is not) (Fine 1996).

Organisational stereotypes (or organisational myths and archetypes) also inform about what group value (or indeed worry about). This idea of archetypes also fits with Dorothy Holland's account of social identities. Holland draws on Geertz's (1983) webs of meaning to describe cultures and identities made up of what she calls figured worlds.

> Figured worlds take shape within and grant shape
> to the co-production of activities, discourses, perfor-
> mances and artefacts. A figured world is peopled by
> the figures, characters, and types who carry out its
> tasks and who also have styles of interacting within
> distinguishable perspectives on, and orientation
> toward it. (Holland et al. 1998, p. 51)

Within such cultural productions 'significance is assigned to certain acts and particular outcomes are valued over others ... these collective "as-if" worlds are socio-historic, contrived interpretations or imaginations that mediate behaviour' (Holland et al. 1998, p. 52).

Organisational culture and organising for safety-critical work

Under these circumstances, workplace cultures can play a particularly important role in organisations, especially safety-critical ones, not only because certain cultures can enhance an organisation's capacity to learn from mistakes by enabling group understanding of shared meanings but also because they work to standardise behaviour. This is because shared values and beliefs allow individuals to conceptualise issues in a stan-dard way that is shared through group norms of behaviour. According to Wheelan (1994), 'agreement about values, norms, and ideologies reduces member anxiety and increases the ability to predict and understand the events that occur' (p. 27).

However, these norms of behaviour can also exert conformity and, hence, deter learning (Hendry 1996) because they focus attention on valu-ing certain kinds of practices and not others and can also inhibit the ques-tioning of those practices. Weick called this 'socially organized forgetting'

(Weick 2012). In this sense, expressions of attitudes, values and beliefs reproduce the status quo as they serve 'not only to orient that individual to that particular social object, but also to position that social object, as well as justify and reproduce, the social system which produced those social positions' (Augoustinos and Walker 1995, p. 29).

In this respect, collectively held beliefs can also be in conflict with one another, especially in periods of organisational change when cultures are contested. Therefore, understanding organisational cultures can provide insight into what individuals in a group value and what kinds of organisational changes are likely to be supported or resisted.

However, as was discussed earlier, as part of the sociocultural perspectives of learning theory, what happens if these typical ways of thinking found within cultures also need to change? Quite often within organisational culture literature, there is an implicit assumption that the existing community of practice or culture is appropriate and desired and 'enculturation' is needed for newcomers to understand existing ways of working. However, these cultures may also lead to conformity and unlearning (Hendry 2004), something identified as inhibiting the development of practices associated with continuous inquiry. Cultures can enhance learning, though they may also sustain existing patterns of belief, and thereby learning to conformity or non-learning (Balthazard et al. 2006). That is, culture may reproduce existing relations rather than change.

Collectively held beliefs can be in conflict with one another, especially in periods of organisational change when cultures are contested.

In terms of the elements of organisational structure and culture found within various perspectives of the organisational literature, the following arising from the literature on structure and culture are important here:

- Processes of differentiation (e.g. job roles and tasks) and integration (e.g. teams) that are used to manage organisational complexity.
- Ways in which work activity is structured through formalisation (i.e. through rules, policies and procedures that constrain the choices of members).
- Ways in which physical layout and tools (e.g. technologies) can structure how work activity is undertaken.
- Distribution of power and authority within the organisation.
- Kinds of groups and/or individuals that identify with the organisation, both inside and outside, and their impact on work activity.
- Informal language and story-telling used to describe work activity.

Purpose of this book

This book provides a sociocultural analysis of the ways in which air traffic controllers formally and informally learn about their work, and the active role that cultures play in shaping interpretation, meaning and collectively held values within work organisation. In particular, it illustrates the significant role that history of working experience has played in shaping what is valued by controllers about their work and the role of organisational cultures in enabling or constraining conscious inquiry.

The book advances research in the field in four ways. First, it moves the investigation of learning in the workplace beyond formalised training programs. The premise of the book is that informal learning is just as important in shaping what people know and value about their work and that this area if frequently overlooked.

Second, it extends understanding of what enhances and inhibits learning beyond the individual – to the analysis of the role of organisational structures and cultures in formal and informal learning. By using an interpretive research approach the book highlights the ways in which the social structure of work organisation, culture and history interweaves with learning and working to guide and shape what is regarded by controllers as what is important and what is not. The book demonstrates how this social construction is quite different from top-down corporate culture understandings.

Third, this study sheds light on the impact of the kinds of structural reform that have been occurring in work organisations around the world and the impacts those reforms have on working people. Technological and organisational reform is leading to changes in work practice and to changes in relationships between workers within the organisation and these have implications for anyone wishing to understand the dynamics of organisational life. As such, the study provides insights into many of the changes that are occurring in the nature of work.

Finally, this conceptualisation of learning and change within the ATC is unique because it is the first time attention has been given to investigating context and learning as a key factor in linking individual, group and organisational performance. Research into learning in ATC has centred largely on cognitive individual performance, performance within teams or more recently on performance at a systems level. By tracing the role of context in shaping formal and informal learning the book shows why interventions at these levels sometimes fail.

Rest of the book

The next chapter outlines some of the technological and structural changes that have occurred since the inception of civil aviation in Australia. In doing so it will set up the stage furniture to later discuss and analyse the ways in which these ways of organising have shaped cultures and identities which in turn have influenced opportunities for learning. Chapter 3 outlines some of the cultures that have developed over time and in relation to how the work is organised. Chapter 4 outlines a framework for thinking about learning in ATC. Two forms of learning- as formalised accredited learning as well as informal learning are introduced as these are both important in high-3 work contexts.

Chapter 5 will discuss how the way work is organised shapes opportunities for experience in certain ways. Four dimensions of experiencing work are outlined (corporeally, cognitively, affectively and socially). These are particularly important in High-3 work. Work may be intense on the body; easy or demanding on the mind; an expression of, or alienation of, the self and it always occurs in a social environment, although some forms of work organisation provide more emphasis of the 'social' than others. The way work is organised shapes opportunities for experience in certain ways and that these organisational structures influence the transformation of experience into learning by enabling or constraining the transition to the next moment in the cycle – reflection.

Chapter 6 will demonstrate the link between individual and collective processes of both reflection and the next phase of the learning cycle, conceptualisation (or making sense of reflections on experience). Both reflecting and conceptualising are emphasised in psychological and sociocultural theories of learning, though as expected, psychological theories of learning emphasise these processes as they occur for individuals and sociocultural perspectives emphasise these processes as they relate to the social context.

It will be shown in Chapter 7 that the contextual elements of organisational culture (such as collectively held beliefs, values and norms) are resources used in the conceptualisation process because they influence perceptual selectivity and the development of individual and collective schemas about work practice. In addition to influencing events, collectively held beliefs and values also will be reproduced by individuals and groups in the kinds of continuous learning strategies used in work activity. Collective beliefs and values, therefore, will lead to some things being noticed rather than others. Thus, schemas enable and constrain both individual and collective opportunities for learning.

Chapter 8 will show how organisational contexts enable and constrain opportunities for experimentation for both individuals and groups. Structures such as teamwork, for example, increase possibilities for experimentation because they enable a shared continuity of experience to occur across team-members and thereby enable the experience to be used as a resource for inquiry to generate increased possibilities for action.

The final chapter discusses the implications for both work design and facilitators of learning in High-3 workplaces. Strategies for both organisational designers and facilitators of workplace learning are offered. The implications for workplace learning of changing organisational structures and cultures are also considered.

chapter two

Structure and change in air traffic control

Aviation in Australia

There were 60.94 million passengers carried on Australian domestic commercial aviation (including charter operations) for the year ending June 2016, an increase of 1.8% compared with the year ending June 2015 (BITRE 2016a). In addition, International scheduled passenger traffic in 2015–16 was 36.229 million compared to 33.865 million in 2014–15 – an increase of 7.0% (BITRE 2016b). Over the same period International scheduled air freight traffic increased by 6.0% compared to 2014–15 to 996,650 tonnes in 2015–16. While inbound freight decreased (5.2% to 490,863 tonnes), outbound freight increased by 19.9% to 505,787 tonnes. In short, aviation provides a significant role in Australia's economy (BITRE 2016b).

Practice of ATC

Within Australia, there are approximately 1000 air traffic controllers, managing air traffic from two air traffic services centres in Melbourne and Brisbane, four terminal control units and 29 towers at international and regional airports (Airservices Australia Annual Report 2016). The two air traffic services centres control for most of Australian airspace: The Brisbane centre controls traffic in the Northern Flight Information Region and the Melbourne centre the Southern Flight Information Region.

The volume of airspace controlled by ATC in Australia covers approximately 11% of the world's surface, or 15.6 million square nautical miles (MacPhee 1992). In comparison, the US airspace covers 20 million square miles and the Canadian 5.3 million. Like other countries, the Australian airspace includes vast oceanic areas; however, unlike other countries, it also includes vast areas that are sparsely populated.

The goal of ATC work and the tasks of air traffic controllers are to maintain separation between aircraft in a way that is safe and allows for expeditious flow of air traffic. Air traffic controllers both direct the flow of traffic and provide in-flight information to assist aircrew in the operation of their aircraft. The organisation of ATC work is divided into the phases of the flight. This division is shared between:

- *The Tower*, which provides airport control and surface movement control, and
- *The Area Approach Control Centre* (AACC), which provides Approach control (responsible for aircraft approaching and departing the airport), Area or 'Enroute' control (aircraft travelling to and from their destination), and 'Arrivals' control (preparing for landing).

Although the controller is responsible for ensuring the safe conduct of the flight throughout that controller's airspace of responsibility, collaboration occurs between the pilot and controller and with other controllers to ensure the traffic moves in a way that is orderly and expeditious. Controllers cooperate with pilots in an attempt to not inconvenience them with delays or by keeping aircraft at undesired altitude levels, and so to assist pilots to achieve their intended flight plan and, where possible, to give them the most direct route appropriate to their flight plan.

Even though ATC practice involves applying just three standards or rules of separation (vertical, lateral and longitudinal), the work is complex because of a range of other factors. For example, the weather influences the traffic flow. Aircraft may be required to divert from an original flight plan due to poor weather or in a desire to get above or below poor weather conditions, aircraft may request flight level changes. Cross-winds or tail winds may alter an aircraft's performance resulting in the aircraft not performing as anticipated. Other environmental conditions (e.g. bushfires and fog) can also alter the flight plan.

The performance of each aircraft can also vary. Fifty-nine international scheduled airlines operated services to/from Australia during the last financial year (2015–16) (BITRE 2016b). In addition to domestic and regional airlines operating in Australia there is a strong General Aviation community consisting of aircraft used for agricultural work, private business and leisure.

Each type of aircraft has its own 'performance profile'. Some go faster than others and the performance of the aircraft can vary depending on the altitude of the aircraft. Some perform sluggishly at low altitudes, while others can climb quickly until reaching a certain level and then the performance changes. The performance profile of the aircraft is influenced also by factors such as which company owns and maintains aircraft; how pilots 'drive' the aircraft; whether the aircraft has a full contingent of passengers; how much fuel the aircraft is carrying and the duration of the flight from departure to its destination.

Place called the room

The main work centre for controlling air traffic occurs from a place known colloquially as 'The Room'. The configurations of ATC consoles within The Room reflect the organisation of airspace sectors across the relevant portion of Australia. There is a hum as controllers talk with pilots and other controllers. Nearby are refreshment facilities, locker space and a recreation area. Lighting is subdued to allow easy reading of data on consoles.

Outward bound flights emanate from the Approach sectors (in collaboration with controllers in the Tower outside The Room) and then aircraft are handed off (i.e. transferred) to controllers operating other consoles. Air traffic controllers might be working in concert with others in the same room, or at other centres or in another country. Controllers might also hand-off control of a flight to military air traffic controllers operating military airspace within Australia. Military airspace is generally active only during RAAF exercises and, thus, may be active only temporarily. Inward bound flights reverse the trajectory of information flow across The Room by travelling from the Enroute sectors, through 'inner' Arrivals and onto Approach, with the work mediated by the 'flow' controller (a senior controller, who assesses the overall traffic flow and decides the Arrivals sequence and conveys this information to the relevant Enroute and Arrivals controller).

Air traffic controllers work interdependently and also facilitate collaboration from others such as pilots to achieve successful outcomes. When the work gets busy, it impacts on others, including if one controller wants to speed up or slow down the pace of the work. The demands of the work will also vary depending on the kind of sector involved. The closer the sector is to the airport, the less time the aircraft spends in the sector. However, because of a range of factors (e.g. aircraft profiles and weather conditions) all sectors can be busy depending on the various problems presented.

The claims in this book are that these work practices have developed over time in interaction with technological developments in that history, changes have brought with them changes in both practice and culture as controllers make meaning of their work. The rest of this chapter will outline some of the main changes to provide a background context for the rest of the book and the discussion about rethinking learning culture and change.

Australian ATC commenced following the end of the World War II. At the time some of the interviews reported in this book commenced, the more experienced controllers spoke of their training under the supervision of personnel who were working during this time. When interviewed those personnel had been in ATC for 30–40 years and could also discuss the changes that had occurred during their time.

Table 2.1 Features of ATC developmental trajectories

Historical trajectory and time period	Features	Aspects within workplace cultures
Inception of civil aviation 1940s–50s	• Recruitment from military • Jet age • Limited/unreliable Navaids • Inquiries into major Crashes = 'care for pilot'	Language as play and war War stories Pilot orientation
Changes in physical sentience 1950s–70s	• Physically moving 'shrimp boats' • Radar + improvements • Keyboard input • AUSCATS (increased info) • Organisational restructuring • Changes in recruitment patterns (no longer ex-military)	Valuing of 'performance', ability and confidence Having 'the picture' Approach is the peak (Hardest/fastest)
Structural reform 1970s–90s	• Facsimile (closure flight ops.) • Economic rationalism (closure of outstations) • Flattening of org. hierarchies (teams)	Masculinist cultures Gun controllers/ adrenalin junkies
Symbolic analysis and integration 1990s–current[+]	• Changes in the bandwidth of experience • Consolidation of airspace • TAAATSs (Mix real-time displays + computer projections) • Inc remote surveillance (ADS, TCAS)	Professional service orientation Safety assurance System collaboration

Continuity and change

Four periods of major technological and structural change are identified with the ways of organising summarised in Table 2.1. Time periods are indicative only and overlap. Importantly, the influence of cultures for all of the time periods can be traced through the history of ATC and remnants still found today.

Inception of civil aviation

Civil aviation in Australia grew out of the activity of military pilots in World War II. At the time only ex-pilots were recruited as air traffic controllers. Implicitly they shared an understanding of aviation as well as a shared history of experience. At this time there was no radar in civil aviation and the labour of individual ATC was divided between three people: a controller undertaking radio-telephony; another who served as an interface between the ATC office and pilots who would physically present themselves and their flight plans, and a third who would write and keep track of estimates on a blackboard.

For example, Figure 2.1 shows an illustration of the kinds of equipment and work organisation of air traffic controllers during this time. The uniformed men at the counter are likely pilots lodging their flight plans.

Historically, the experience of ATC work has been subject to a number of significant changes since its inception, and these changes were arguably just as demanding on those controllers, when they were first introduced, as the changes being implemented now and planned for the future.

> We had a number of nasty airline accidents in Australia in the late 1930s. Because there were no navaids and things were incredibly crude. These accidents, investigations and royal commissions that followed, resulted in navigational aids, air traffic control being founded and everything else. Then just after WWII, we had towers and HF radio and flags and it was pretty crude. You had the war time system adapted to peace time. But once aircraft were flying on the main routes you had very crude Enroute control, messages being passed on bits of paper through the flight service operator talking to

Figure 2.1 Area control inception. (Credit: Airservices Australia/Civil Aviation Safety Authority. Photo courtesy of the Civil Aviation Historical Society of Australia.)

the planes on HF. All that hashy radio that we [air traffic controllers] wouldn't lower ourselves to listen to, so the poor wretches in-flight service, talked to them all and passed bits of paper back and forth through a hole in the wall. I've seen the photos of it. A hand appeared - and the poor little wretch with the head phones on, listening to radio Peking, and everything all in this hash, would pass your 'climb to 8000' message to the DC3 between Sydney and Brisbane. And you didn't lower yourself to talk to him.

There were a number of major accidents in '48 and '49 Very bad ones and there was a paranoia in the public about flying. After all this wonderful flying in the war, so comparatively safe, now people were being killed all over Australia. The Royal Commission [into one of the accidents] found that air traffic controllers were plotting him [the pilot] and separating him, but nobody was looking after him. Nobody was actually checking to see if the navaids worked. / Anyway, the decree was made that no-one looked after aeroplanes after they got in the air. We just separated them, but, did he have enough fuel on board? – that sort of thing – so operational control was born.

The number of fatal air crashes referred to in the quotation above, and the concomitant outcomes of the various inquiries that resulted from them, significantly influenced the emergence of the organisation of work within ATC. The pace of the work was mediated by the rudimentary nature of technologies in use, which included the aircraft. A flight from Melbourne to Perth, for example, took seven hours (compared to the current less than three) and occurred at night to spare the passengers the inevitable air sickness caused by the bouncing around at 7000 feet in the thermal air currents rising from the Australian desert.

Many older controllers, argued that in many respects the work was more difficult and potentially more complex in the past because of the crude forms of technologies available. The pace of the work also changed dramatically, with the introduction of jet technology, as the following controller explains.

Then, [the 1950s and 1960s] the first jets had arrived and it just broke the back of air traffic controllers around the world. The jets were doing three times the speed of all the piston engine airliners. They were higher, faster, descending into the others. All the jets had international gear which couldn't work the Australian navaids. So you couldn't get distances from them, you couldn't get anything from them. They couldn't give you any information. So you just had to *'Keep vertical, keep vertical'*. But everyone was obsessed by how expensive these things were to keep in the air. The operating costs of jets are 10 times that of the old DC plodding along. So you had to get them through. So they did these incredible things, / The sort of things they used to do was: the jet would come over the top of the 2 or 3 slower piston engine aircraft and then, keeping vertical, get the jet to do a 180, turn around, sight them, this was by day or night, come at them from the other direction, this is 10 miles from the airport, sight them going underneath him, then descend through them, get below them and come around and pass them underneath and get in front of them again. And the manipulation and the workload and the talking and being positive *'You know what you're doing'* in something like that. And that was a standard ploy to get the jets through the other traffic. Over the top, 180, sight them, descend through them, back underneath and get in front of them by the runway. And all of this sort of thing was going on all the time, we don't know we're alive! Because we've [now] got all this gear.

In addition, the arrival of the jet age sped up the temporal pace of the work of the controller, which had significant ramifications for those working at the time, as the following interviewee discussed.

A great War story, and true. He'd had a morning shift, followed by a doggo. And he'd done the morning shift in Sydney tower in 1957. He went home and had a sleep but was due at a friend's place for dinner that night. He didn't want to go. He wanted to sleep because he was back in the tower at 11 that night. They were committed to going out to dinner and when he got to his friends place, somewhere around Bondi or somewhere, and the aircraft were coming over the top all the time. And he was sitting there

> listening to these aeroplanes going over and he
> made several references to *'those bloody aircraft'* and
> that he'd just had enough of *'those bloody aircraft'*.
> And after dinner he got extremely tired and he lay
> down on the sofa to have a rest. He was going to
> drive straight to work following dinner. And he had
> this breakdown where he couldn't move. Couldn't
> get up. Just couldn't face it anymore.

The quotation above is the account provided of the war story the 'break-down of the labouring body'. In the war story, the above change in the temporal pace of the work wore the controller's mind and body down, such that he was no longer able to continue.

Changes in physical sentience

Technological developments associated with the arrival of the jet age included the introduction of basic radar. On the sectors that had radar, the first generation of radar display provided a very basic set of history blips and controllers would manually write the call signs and flight level on small pieces of paper (called shrimp boats) which they would physically move across the screen in accordance with the path of the flight (Figures 2.2 and 2.3).

Figure 2.2 Melbourne AACC bright display – early days. (Credit: Photo courtesy of the Civil Aviation Historical Society of Australia/Barry Kemp collection.)

Figure 2.3 Bright display showing controller moving shrimp boats. (Credit: Photo courtesy of the Airservices Australia/Civil Aviation Safety Authority Civil Aviation Historical Society of Australia collection.)

However, these innovations were sometimes unreliable. For example, instead of tracking a low flying aircraft on take-off, the radar beacon would sometimes pick up a fast moving truck driving over a bridge near the airport, resulting in considerable stress for the controller involved! War stories of the time included the time the morning tea was delivered by a lady wheeling a large tea trolley into the centre and on one day, she crashed the trolley into the back of a console, sending all the shrimp boats tumbling to the floor.

Development of the radar data interface

The post-war population boom and a growing confidence led to a proliferation of airports across Australia.

The pool of ex-military pilots diminished and controllers were sourced from the general public with the exception that they needed to have a background in aviation (ex-RAAF, to hold a pilot's licence or to have a parent who was part of the industry).

The division of labour reflected the bureaucratic processes of the day and seven levels of ATC expertise were developed. These reflected the level of congestion and the pace of each sector. The lowest level represented a journeyman controller, who began on Enroute sectors (regarded as slower and simpler and a necessary part of screening for future positions, since to become expert, controllers needed to develop a mental picture of the airspace), arrivals – having more aircraft and changing flight levels was higher with approach control being the highest and best paid of the levels. In addition controllers at a centre such as Sydney would have been paid more than someone working, for example, in Brisbane.

Technological improvements in radar displays resulted in flight and aircraft details now being available and connected to the blip trail of the flight path on the console (Figure 2.3). This change resulted in less explicit communication required between controller and pilot. Controller input for radar consoles now occurred through a keyboard. These computer-mediated radar screens went through a number of iterations (called 'ATCARDS' and 'AUSCATS') (Figure 2.4).

Procedural ATC on non-radar sectors still involved writing flight progress, on to plastic flight strips that were positioned below a large map of the airspace. As the aircraft progressed through the airspace, the flight strip was manually updated and moved in accordance with its flight path across the sector.

In terms of work practice, these technological innovations resulted in two significant changes: the controller no longer needed to physically move the aircraft, pointing out where the aircraft was going to be, and the reduction in the amount of communication needed between controller(s) and pilots, since the information was easily available and did not have to be requested (Figure 2.5). These technological improvements have since

Figure 2.4 Melbourne AACC bright display – 1970s. (Credit: Photo courtesy of the Airservices Australia/Civil Aviation Safety Authority Civil Aviation Historical Society of Australia collection.)

Figure 2.5 Melbourne AACC ATCARDS/AUSCATS – 1994. (Credit: Photo courtesy of the Airservices Australia/Civil Aviation Safety Authority Civil Aviation Historical Society of Australia collection.)

had some unintended consequences for those involved in the accredited learning process, as the following controller explains.

R: One of the things we say with the new radar is that it makes it harder to know when they have 'got it' now because now we have got the altitude readout of speeds etc., — you can't see what they are thinking. Before they had to ask. They were always asking aircraft for their speed or an altitude and you could see what they were thinking. Now, you are just sitting there and you don't know what they are thinking. You don't know if he got it right by luck and he didn't even see the problem, or if he actually knew what he was doing. The old radar was good like that. A piece of basic radar and you had to work. / You don't know as much about what people are doing. Instead of trying to work out what they are thinking. That is why at times you say to them *'What were you going to do here?'* just to find out if they are thinking about it. If you get a dumb answer or a blank look, you know he wasn't thinking about it. If they can come back quickly you know that he has seen it.

I: Whereas before, you could detect by the questions that the trainee was asking?

R: Yes. Things he was saying to pilots and what he was doing. You could think *'Oh yes. He has seen that, he is doing that'*. Sometimes you couldn't, but often, it was far more obvious what they were thinking. When you are looking at a situation going wrong, it is really nice to know whether or not he is thinking what is going wrong or is he thinking about something else and hasn't seen this is going to go wrong.

The advances of technology in this case make the thinking of the trainee opaque to the instructor, who then has to use other methods to determine the trainee's planning, thinking and decision-making.

Organisational restructuring

As discussed earlier, ATC work had been organised across three roles or functions: The air traffic controller who tracked and separated the aircraft, a flight service officer who communicated the controller's instructions to the aircraft via radio and a flight operations officer who liaised with the pilot and checked that the aircraft was properly equipped before departure. These divisions and job roles continued for just over 30 years.

The flight operations department was in place from the early 1950's until it closed in 1991 as part of a policy to reduce the labour force and associated cost. The introduction of electronic forms of transmission such as the facsimile machine resulted in online input of flight plans, and enabled closure of the flight-operations departments and the requirement for face to face contact with pilots.

Traditionally, this was the place where pilots (and their representatives) would either contact or visit to lodge their flight plans. The technological capacity developed meant that these could now be lodged automatically. The following quotation, describing the impact of the closure of this part of the ATC work organisation:

> At the stage that I first went to the [ATC Centre] we were running briefing offices at every general aviation aerodrome around the country. The purpose of the briefing offices [was] two fold. One was the provision of flight information for pilots like normal flight planning procedures. The second one was/ pilot education. It provided a link between air traffic services, air traffic controller flight service and the pilot. So that the pilot could go out, if flying around, [and] maybe not clear about something when they came back, they'd call in the briefing office and have someone clarify. Or maybe they did something wrong. Instead of a rather impersonal sort of phone

> call of whatever, it was quite common to ask a pilot
> to report to the briefing office so he could sit there
> and he could actually go through what the problem
> was and possible ways that they could solve that
> problem. It [was] an education tool. There's a huge
> advantage to this. That was our main interface with
> general aviation. We closed all our briefing offices
> around the country. / But immediately there was
> a noticeable difference to me. You immediately lost
> that contact, face to face contact you had with pilots.
> Because everyone worked the briefing roster. You'd
> be over there, you might have to do one shift in 5 or
> 6 or something. But at least if you're in the briefing
> office you get to know the pilots as they're coming
> in, you talk to them. They know you, they recognise
> voices when you're talking to them on the radio. So
> that went. It was a great loss to us.

The division of labour also underwent significant restructuring as part of national and international economic reform agendas. The previously developed seven layers of ATC hierarchy were flattened to three levels and integrative mechanisms of multi-tasking were introduced as part of the industrial relations negotiations and agreements. These involved controllers gaining and maintaining currency to operate three sectors, enabling workforce flexibility. In addition, team-based working practices were introduced.

Changes in recruitment

Changes in technology and work organisation and changes in recruitment practice have also influenced the level of aviation background of controllers. When previous ex-war and aviation familiar cohorts dried up a new approach was taken.

Australia's first 'professional' air traffic controllers (where there was no formal requirement to have a background in aviation) occurred during the late 1960s. Between the 70 s and 90 s those without an aviation background undertook a two year training period. Yet, many controllers and trainees still had affiliation with the aviation community, in part because the career trajectory for many during this period involved the controller being sent to an 'outstation', where he may join the local aero club.

Changes in recruitment practice changed the nature of the experience controllers can draw on to understand the implications of work activity for the broader aviation community. The following excerpt illustrates the perceived difference between those controllers with many years of experience who have been recruited from 'within' the aviation sector and those who, as a result of changed recruitment policy, have joined more recently.

> It's a funny culture now. When I first started off in ATC, and I started off with the air force, and then I transferred over.... To actually do the job, you had to have some, you had to have an interest in aviation. You used to have to love aviation and love aeroplanes. Nowadays, people do it because they saw an ad in the paper and they say *'Oh yeah, I'd like to do that'* and they go in and they pass the psych tests and the aptitude tests because they're bright and pretty brainy, and they do their course, and they come out and they do their rating and they're at the window and they see an aircraft taxi past and they say *'Is that a 737 or a 767'* or something like that, and they don't know the difference between aircraft, because they had no requirement to have any interest in aviation.

These changes resulted in controllers entering the aviation system with less background knowledge about the aspects of the aviation system than their previous counterparts with the consequence that they have been on a much steeper learning curve than would have been the case for their predecessors with a background in aviation. As will be discussed in later chapters in the book, these changes are coinciding with a decline in opportunities to informally learn about the aviation system.

Structural reform

Although the flight operations department had closed, the roles of air traffic controller and flight service officer continued until the early 1990s, when these divisions of labour were also dismantled. Historically, the two kinds of operational work (ATC and flight service information) both required licensing or ratings for individuals to be able to do the work. The two streams had their own training and rating procedures, and up until the changes in industrial relations policy leading to award restructuring, staff rarely 'crossed the line' between the two occupations. Within these two aspects of operational work (e.g. ATC and flight service), there have been divisions that have deep historical roots. These divisions created subgroups based on the kind of work undertaken and, as the next transcript shows, extended to how each group entered, for example, a building at a particular Centre and where they parked their cars.

It had always been an 'us and them' mentality, particularly in large Centres where there is a physical division between the areas. I will give you an example, in Brisbane, Flight Service was one end of the hall, ATC was the other end of the hall there was a wall between us. We all opened on to the same hallway but it was physically at the stage where Flight Service Officers would use one entrance to the building and ATC would use the other entrance to the building and didn't talk to each other in the car park, didn't know each other personally.

According to the above respondent, in this particular Centre, when the wall physically separating the two groups was removed, each group acted like 'fish in a bowl', each working within their own space up to close to where the wall once was but remaining within their own 'territory'.

The formal division of labour between ATC and flight service changed with the commencement of organisational downsizing, supported by policies of award restructuring and technological enhancement. As the respondent documented in the transcript above continues, flight service officers perceived that there was an underlying political agenda behind the offers of re-training made when the flight service position became redundant.

Well it's — getting back to the attitude of some — some came in [to do the conversion course] with the attitude that 'This is just a circuitous route for the [organisation] and a bunch of air traffic controllers to get rid of us / because they don't want us anyway, getting rid of our job and [they] can't just get rid of us with involuntary redundancies without offering us this chance [to do ATC conversion training] and they are going to make it as hard as possible for me.

During the period of award restructuring the suspicion and concern indicated above was reported in all centres, but not by all staff. Some staff who had undertaken and successfully completed the conversion course believed that the animosity was history — part of an old culture and that it no longer existed. As the following ex-Flight Service Officer, now controller, explains:

I know I came from that – the old school in Flight Service was that ATC is that Out There and we were very much 'us' and 'them'. And so you start out

your training program with the feeling that perhaps
you're not wanted. So that's one of the major things
that I'm interested in – is to make sure that that
kind of barrier is broken down. And I've quickly
discovered that it doesn't exist — that they're all
just human beings and they've actually got a really
good attitude towards training.

Symbolic analysis and integration

The fourth and final period of technological development and structural
change spans from the beginning of the twenty-first century to the present
day. Technological developments and economic restructuring resulted in
centralisation of services and consolidation of Australian airspace to the
two major centres (Northern and Southern Flight Information Regions)
with the subsequent closure of outstations. Economic reform also resulted
in an increased streamlining of ATC training, removing nice to know job
experiences such as observational attachments to airlines.

Changes to the bandwidth of experience

Rather than experience in previous ATC sectors being the requirement for
movement onto specialist sectors such as tower and approach, streaming
for new recruits was introduced in line with international practice.

In addition to changes in recruitment practice, there have also been
changes in the duration and content of pre-workplace training programs.
Economic and political pressures exerted on the organisation in the 1990s
led to a streamlining of course content with removal of 'nice to know'
material, with a subsequent reduction in time taken in training. It is contended that these changes led to controllers being competent in job tasks
and job performance, but who have less understanding than their older,
more experienced counterparts, of the relationship between the controller's role and the roles others play within a complex aviation system. The
following quotation from an 'old hand' summarises concerns held about
the level of background aviation knowledge held by these controllers.

We did a two year course and went out into the field
and we did airline attachments and we did outstation attachments and we went to RAAF attachments and all sorts of things. We got to know about
aviation. I already knew it because I had a pilot's
licence before I joined. You got right into it and
you met the pilots and you flew in the aeroplanes
and you checked out the [aircraft] profiles and all
sorts of things. There are guys [in The Room] now

who have never even been inside the control tower. They have come out of the College and / they have never been inside a cockpit of a jet or anything else like that. They don't know what an aeroplane does. What it looks like. It is just a blip on the screen that goes so many knots. Then you have even got to tell them how fast they go. *'How fast does he really go?'* and you say *'Well he does this, that and the other. The cockpit looks like this inside, visualise what's inside the cockpit. The pilot hasn't got a control panel, just a little stick on the side.' 'Oh has he?'* Just little things like that and get them to visualise what's going on. At the moment it's just a production line.

The narrowing of opportunities to learn about the aviation industry as a whole has also been expressed by others working in the aviation industry in general. Anecdotal evidence suggests that recruitment and training strategies for pilots, for example, have undergone similar changes. In the past, many pilots learned to fly in aero clubs and flying schools whereas the new generation of pilot is being trained in universities. Changes are leading to increased specialisation together with a decline in opportunities to informally learn about particular aspects of the aviation system and one's role within this system.

As traffic volumes have grown there has been a need to introduce procedures aimed at smoothing out the temporal highs and lows within the workload so that peaks and troughs in workflow are avoided. The implementation of procedures – such as Standard Arrival Routes or STARS – makes the work more orderly and routinised. However, because of the highly interdependent nature of ATC, a reduction in complexity and orderliness in one sector has consequences for another sector, as the following Arrivals controller explains.

The Approach still think they are prima donnas and that is it. It has always been the case there. They are trying to make their job easier [to be able to bring an abinitio [trainee] straight in onto Approach. They still haven't realised I think that there is a lot more work load now on Arrivals than there used to be — just clearing aircraft in and descending aircraft. Well now it is a lot more work tidying up [for] all the STARS and making STARS work for them [Approach] as well. All the holding is done in Arrivals now. It used to be done some on outer sectors and some on the Arrivals. But, we [now] do a lot of work.

Abstracting interpretation

The introduction of The Advanced Australian ATC system or TAAATS commenced in 2000, which led to a need to differentiate between real-time data presented and data produced through computer processing (Figures 2.6 and 2.7). Up until this time, the symbols being interpreted had remained a 'real-time' representation of the air traffic pattern.

In the following quotation, the controller is discussing how on some sectors (radar) the controller has on the screen a real-time representation of the traffic pattern in the air. This is in contrast to the situation presented to a colleague nearby working a procedural sector – now called flight data processing – who has the same screen, though the representation is a computer generated image of where the computer (based on the information provided to it) has positioned the aircraft. The visualisation referred to at the beginning of the quote refers to what procedural controllers needed to do previously, when working on a procedural sector (i.e. turn the information presented on the flight strips into an internal visualisation of the traffic pattern).

R: You don't have to visualise up here [head] any more — you can see it all on planned display. But a lot of the areas [i.e., in Enroute] it isn't where the aeroplanes are – it is where the computer thinks they are. / They can, in their heads, think that when they're here they're [using] radar / but it's not a radar as such. So there's huge cultural, human factor type problems involved in that and I think that's kind of exciting.

Figure 2.6 TAAATS/Eurocat console close-up, 2010. (Credit: Photo courtesy of the Civil Aviation Historical Society/Phil Vabre.)

Figure 2.7 TAAATS/Eurocat wide view, 2007. (Credit: Photo courtesy of the Civil Aviation Historical Society/Phil Vabre.)

Under the TAAATS system, computer-mediated Enroute procedural control is no longer undertaken using paper strips on a flight progress board, but instead is undertaken using a computer display, which is a facsimile of a 'real-time' radar display. The display is not a representation of where the aircraft are, according to real-time radar paints, but is instead a computer-generated image of where the computer-based information anticipates the aircraft is or will be. The difference in whether the information can be interpreted as a real-time radar display or a computer generated image will be in the kinds of symbols used on the screen. These changes required an increase in the interpretation of the meaning of data and the development of strategies for data interrogation to determine the reliability and authenticity of that data.

The implementation of systems such as TAAATS has involved a greater degree of abstraction and interpretation than had previously been the case because there are now an increased variety of symbols for the controller to interpret. This is also exacerbated by a greater variability in the standards and procedures to be applied to aircraft with differing technologies (and demonstrated on the computer screen by different symbols). Some aircraft, for example, have their own forms of surveillance and radar detection equipment on board (such as Collision Avoidance Systems), but not all aircraft flying through controlled airspace have these technologies. Aircraft (and airlines) investing in such technologies wish to gain full advantage of such systems, which means a desire for smaller degrees of separation and greater flexibility in terms of, for example, flight path trajectories. In response to demands from the aviation community there are also now differing airspace procedures on some airspace sectors. In some airspace sectors, some aircraft can choose to be fully controlled and separated by the controller, and others can choose to just receive a 'radar advisory service' (similar to that which used to be provided by Flight Service Officers but are now part of the role of the air traffic controller). These differences in aircraft technologies and procedures are indicated on the screen by different symbols. Within TAAATS, there are other aspects of information the controller has to interpret, as the following controller explains:

R: A workplace controller needs to know more about the system and the implications of the system when something he does is not done properly because everything feeds on the system. This 'feeding the elephant' phrase …. If you don't feed the elephant [input data into the automated system] then someone else is getting the wrong idea and is acting on information that is wrong, but the computer thinks is right because no-one has told it otherwise.

I: So that's system awareness of the system and implications of the system?

R: Yeah because it's central. Everything around TAAATS works around this system and they continually cross check each other the [inaudible] data thing and the flight data thing. They're always linked; everything's linking in and updating everything else automatically. So I could change something on a strip and that'll flow through for the rest of the flights and alarms and bells won't happen at the other consoles to say *'this has changed'.*

Signalled in the controller quotations above is the issue that computer-mediated interdependent work requires a greater need to recognise the impacts of one's actions on others and to interpret data-driven information from the perspective of the totality of the whole system. In the past work that influenced the work of someone else would require contact through direct communication.

Technological changes continue in this complex sociotechnical system. Airservices Australia have an ambitious program to collaborate with the Australian Defence Force and to bring all airspace into one system. This will involve integrating civilian and military ATC systems so that all airspace will be harmonised and part of one system by 2021.

Conclusion

Historically, ATC work has moved from one of physically acting on the tools available to one of interpreting the meaning of data and symbols. The move from shrimp boats to a computer-mediated display has changed the relationship between thinking and action for controllers because these cognitive processes are now mediated by symbols and data available on a screen.

As technological innovations have been made, they have enabled a number of other organisational structures to change, which have in turn modified the nature of ATC work and how personnel make meaning of what they do. The impacts of these changes and their interaction with the practices of continuous inquiry and learning will now be discussed.

It has been contended by Zuboff (1988) that technological change impacts on work practice by restructuring the work environment and, in so doing, abstracts thought from action. As technological improvements are implemented in the ATC workplace, it has made more information available to the controller, and through abstracting such information, the work is increasing in complexity.

Just as technological change increased the temporal flow of the work since the inception of civilian ATC, the technological developments occurring at the beginning of the Twenty-first century look set to continue increasing the intensity and pace of work. The changes discussed above also signal an emerging complexity as the work increases in its reliance on interpretation of symbols and abstraction of information. Together with these changes, have come shifts in the ways in which people acquire knowledge informally through work activity in ATC. Significant changes in the nature of work are resulting from the introduction of more sophisticated technologies. Technological change is leading to a tighter coupling of job roles within the ATC system. The trajectory of these changes lead to the conclusion that the future of ATC work in particular, and High-3 work more generally, is likely to have the following characteristics:

- Be more temporally demanding, though more regulated
- Be more complex in terms of interpreting symbolic information as a result of the variety of data available
- Lead to shifts in the ways in which individuals can gain their needed background knowledge
- A decline on the reliance of sentient know-how
- Increase the demand for the development of intellective skills (skills in data reading and interpretation)
- Increase the demand for systemic awareness and understanding of the impacts of work roles within tightly-coupled, computer-mediated information systems
- Lead to stronger ties to teams but with weaker ties to broader work or occupational groups

The key question, then, is how people working in these environments informally learn about their work as well as their place within the system or 'machine'? Once again, in order to look ahead, it is useful to consider that people over this period of time have informally learned about their work practice and to reflect on the implications of these changes for the future.

chapter three

Cultures within ATC work

Role of culture in organisations

Organisational culture is a set of understandings shared by a group of people that are largely tacit among members. Values, beliefs, attitudes and norms are used by members to justify certain decisions and behaviour. Beliefs become values when they are applied to evaluate someone's work quality. Thus, it is contended that what is regarded as 'good performance' is always relational – it is always linked to how well one's performance is in relation to the performance of others. This is important in the sense that what constitutes 'good' performance is always set up, to some degree, by the group and not by external agents.

Analysis of organisational cultures will yield important insights into collectively held beliefs, values, history and experience of workplace learning and change, and the common context that links individuals to groups. Understanding something of a group's 'culture' will offer some insights into what that group has learned (and what it has not); what it regards as important (and what is not); and what is regarded as learnable (and what is not).

A 'group' is here defined as

> a collection of individuals (1) who have significantly interdependent relations with each other, (2) who perceive themselves as a group, reliably distinguishing members from non-members, (3) whose group identity is recognized by non-members, (4) who, as group members acting alone or in concert, have significantly interdependent relations with other groups and (5) whose roles in the group are, therefore, a function of expectations from themselves, from other group members and from non-members. (Hartley 1996, p. 401)

This definition highlights the structural features of groups (e.g. their interdependence) and alludes to the cultural features that give a group its identity (e.g. the norms of anticipated behaviour). Examining evidence of organisational culture also provides insight into the way practices are embedded in organisational structures because it highlights the relationships and interdependencies between groups and the types of accounting systems that individuals and groups use to explain their actions.

Cultures are demonstrated in the way people communicate understanding, their shared norms of behaving, the stories they tell and the stereotypes they use to account for group membership. In this respect groups become, as Loui (1986) suggests, 'culture bearing milieus'.

In this chapter, aspects of group membership discussed in the interviews will be reviewed. These include gender-related groups, broad occupational groups and groups based on tasks and roles undertaken. Some of the communication patterns and norms of behaviour will also be discussed in terms of what they reveal about collectively held beliefs and values within ATC.

Collectively held values and beliefs are shared also through narrative as 'war stories'. Group norms provide the foundations for the development of stereotypes of behaviour and also become used in language to portray archetypes of behaviour (known as 'Gun Controllers' and 'Adrenalin Junkies'). It will be argued that different experiences of work groups are leading to different norms and these have implications for practices of inquiry. The elements are outlined in Table 3.1.

The rest of the chapter will elaborate on the kinds of workplace cultures evident in ATC.

Group memberships

As in most organisations, people in ATC belong to multiple groups. Some of these groups are work related and some are not. Some groups

Table 3.1 Aspects of workplace cultures

Cultural elements	Attributes	Examples in ATC
Group membership	Language and symbols used to justify membership and non-membership to particular group(s)	Groups: gender, professional affiliation, work groups: operational and non-operational groups, airspace sector groups
Communication patterns	• Collectively held beliefs and values that members share • Informal language used, stories shared (real and imagined)	Importance of ability, performance and confidence Stories based on actual practice – 'war stories', mythological archetypes 'Gun Controllers and Adrenalin Junkies'
Norms of behaviour	Expected ways of behaving	Justifying work in terms of one's performance; putting others down; 'Lone Rangers' and 'Team players'

are external to the organisation, others are within the organisation. When individuals define themselves as part of a group, they justify their behaviour in terms of group norms and develop a group identity in terms of their interactions with others in the group and the responses of others to group behaviour (Fine 1996). For example, individuals identify with their occupational group. An occupational identity occurs, for example, when workers justify the work undertaken by their work group and explain to themselves and others why what they do is admirable and/or necessary.

Some individuals identify with groups that may or may not involve other members of the occupation. For example, some male controllers (and the majority of all Australian controllers were male, and mostly of Anglo-Celtic descent) identify with their gender, their masculinity. Masculinist group cultures have been researched within a number of occupations such as automotive, metal fabrication, engineering and computing (Collinson 1992; Owen 2013). The following controller discusses the ways in which masculinist group cultures would be displayed in The Room.

I: We were talking about the culture of air traffic control.
R: About air traffic in general. I think if you talked about teams — it's very ego driven. The term 'anti-woman' isn't right. Air traffic controllers respect someone who can do the job and that doesn't matter what the sex is, but a lot of them are very much like this group of people where they'll happily have a go at someone and use the feminine joke.
I: What? Like 'you're such a girl'?
R: Yeah. I think nowadays the majority of them don't mean it, but there's still that culture, there's still a very ego driven culture.

For this controller, the term 'very ego-driven' infers a culture that is male-ego driven, as evidenced by the gender-based comments that accompany the explanation. The 'feminine joke', in this instance, is a derisive term and an indicator of masculinist culture. As will be shown later in this book, those identifying with masculinity tend to emphasise behaviours associated with performance and perseverance, with 'individualism' and being 'tough'. It is argued that these collectively held beliefs inhibit inquiry and, hence, learning. Controllers may also identify with groups external to the organisation that have nothing to do with work activity, and as such may seek little involvement in work activities beyond what they are required to do. However, many controllers not only identify with their occupation, but also with others in similar occupations within the aviation industry.

Professional affiliation with the aviation industry

There are controllers who have a strong association with the aviation community and define themselves in terms of this affiliation. These controllers may either own an aircraft or at least operate as a pilot in their spare time and value knowledge about aircraft types and performance. Controllers identifying with the aviation community are likely to have a strong service orientation to pilots, since they have experience in being on 'the other side'. The following quotation illustrates the 'love of aircraft' spoken of by controllers identifying with the aviation community.

> From ever since I was a little boy I lived next to an aerodrome, Essendon Aerodrome just over here, and I've always had a fascination for aeroplanes, and all through my youth I went to air shows and hung over fences and was always fascinated by aeroplanes. When I commenced work, my very first pay that I got I started to learn to fly — at the age of 16. At the age of 16 I got a pilot's licence and I was working at Essendon Airport with the Department of Customs and Excise, to be near aeroplanes, to have an involvement with aeroplanes. But I felt that when I finished school, the qualifications that I had, would not get me into the airlines or the Air Force. I was thinking along those lines. When I started work my first thing that I wanted to do was to learn to fly, so I did that and I got my licence. And then it seemed strange coming to Essendon Airport flying across Melbourne as a 16 year old to collect my pay. I couldn't drive a car, because [you had] to be 18. But it was very clear from after about a year in Customs that really wasn't what I was interested in. It was more a clerical type job. And I saw an ad in the paper one day for ATC [and I applied and got it].

Controllers also identify themselves with others who have similar histories of experience within an aviation-related occupation. For example, when conducting the research, the controllers would be asked to describe their background. They would typically commence with a shorthand description which outlined the kind of training they had undertaken (e.g. 'I did a short course at Henty House' or 'I came off Utol CC Course 4'). This information provides the listener with immediate information about the history of the speaker: the level of aviation background the person had, their historical positioning within the organisational system and the aviation community more generally. If the speaker described having undertaken a 'long' course, for example, then they had no previous aviation experience or background of note, whereas a 'short' course indicated the speaker was likely to be either a pilot or to have had experience in the military. Controllers describing 'CC' course or Conversion Course (and the number) would indicate the speaker was previously in either Flight Service or Airways Data operations, and had undertaken the training to upgrade to an ATC position. Although the situation has now changed, individuals would be offered a course in order of their ranking from the selection tests. Thus, participants in 'Course 1' had received better selection scores than those in 'Course 4'.

Work group identification

The segmentation of tasks and subtasks and the meaning and values controllers place on undertaking these tasks result in controllers identifying with a number of groups within the work activity of ATC. These groups are defined by whether people are licensed to work in The Room (or not) and what kind of work they do (e.g. which sector), at which Centre they work in, and their background in relation to aviation and the organisation. For example, subgroups included: operational and non-operational work groups, work groups based on different types of operational work within ATC and between those operating different sectors. These are outlined in Table 3.2.

Operational and non-operational work groups

Within ATC, one of the ways in which groups define themselves is based on the work those groups perform within the organisation. One distinction that defines the kind of work undertaken is whether it is regarded as 'operational' or 'non-operational'. Operational work within ATC is work that involves the task of separating aeroplanes, whereas non-operational work involves any other work. Operational work is defined by organisational members as that performed in a 'real time', dynamic, environment. Air traffic controllers differentiate their work from that found in other parts of Airservices Australia (or indeed in other organisations), based on the notion that their work cannot be stopped once it commences. Air traffic controllers define themselves as different from other 'non-operational

Table 3.2 Summary of work group cultures

Work group cultures	Examples
Operational/non-operational work	Those licensed to work in 'The Room' and those who are not (e.g. Tr Dept)
Between different types of operational work types	Flight Service Officers/air traffic controllers
Within different types of operational work	Approach/Enroute controllers

staff' because they are the ones who 'earn the money'; they define themselves as being at the 'sharp end' of the organisation. In the following
transcript, for example, the controller is talking about a lack of respect
controllers have for non-operational staff, in part because those staff
members do not seem to appreciate the work controllers perform.

> People [ATCs] wouldn't necessarily have respect for
> managers that came in from outside because there
> is a perception that you have got to understand the
> job. [A CEO had] the insensitivity of saying, back
> like a few Christmases ago, wishing everyone a
> 'Happy Christmas and a nice break away'. Most
> of the employees and the money earners are shift
> workers who would be working over the Christmas
> period. [I: That's one of the busiest periods for you.]
> Exactly! Here was he wishing … he's talking about
> the admin staff, the people who don't earn the
> money, if you want to get down to it.

The operational/non-operational divide was also evident between
those rated controllers in The Room and those (often unrated) ex-control-
lers who work in the training department and now have full-time training roles. Historically, the training department has been variously called
'club vege', 'God's waiting room' and 'the Rest Home for the invalid'. These
labels portray the value that rated controllers place on undertaking the
work of ATC. Those who cannot undertake the work because they do not
have a rating are of limited value (best encapsulated by the saying 'those
who can – do; and those who can't – teach'). In the following transcript,
the non-rated ex-controller explains the perception he believes rated controllers have of the work he and his peers do in the training department.

> It's part of that dilemma. If 'we're the people - we're
> the people delivering the job', [then] *'the rest of you
> are all parasites. Including all you training people.'*/Our
> fellow controllers say, 'Having a jolly down at [the
> College]? *You people in the* [Training Department].
> *You aren't really separating aeroplanes. You're getting
> money for nothing.'*/ It's the old thing about, 'if you
> can't do it — train'. Alright? And everybody believes
> it. I suspect I even believe it.

In the quote above, it is evident that controllers define themselves by the
kind of work they do and one of the first means by which that is defined
is if the work is 'operational' or 'non-operational'. Operational work is

more highly valued than work performed by those who are not licensed to operate within The Room.

Prima donnas *and sector wogs: Approach and the rest of the room*

Earlier the various ATC sectors and their structured work relationships to one another were outlined. This arrangement results, at the Melbourne ATC Centre, for example, in Enroute sectors, being on one side of The Room and Approach being on the other. This physical separation provides a foundation for the development of sub-cultures within The Room is based on work activity and further developed in the physical location of work groups (where Approach controllers were sometimes labelled *prima donnas* and Enroute controllers 'sector wogs'). As will be discussed later in the book, within-ATC beliefs and values about ability, performance and confidence set up sub-groups based around the kind of work undertaken and these beliefs both enable and constrain learning. In the following extract, an Arrivals controller describes his work and how Approach controllers (those traditionally perceived to have the most ability and to work at the highest level of performance) would traditionally behave.

I: People have talked to me about the differences in ATC culture and some of it is really strong and people have talked about *'There is an Approach culture and there is a rest of the room type culture'*. Where do Arrivals fit in that pattern? What's your perception?

R: We probably fit in the middle. / [Historically] you started on outer sectors and quietly built up [expertise] there. Usually went to Tower attachments as co-ord [coordinator] and then back on to maybe Arrivals and you go back to the Tower and do basically ADC [aerodrome control] which is the senior tower controller and then you would probably go to Approach. That was the general [career path] for everybody. People dropped off along the way as they reached their ability or said *'I don't enjoy it any more, I don't want to go any further'*. That was how it was then. There was this little pinnacle up the top there and all these people were going *'Hey, we're the Approach controllers. We know how to do the job and you don't'*. Usually, putting others down — *'Oh it's a piece of piss* [another controller's airspace]. *I could do it ... and we used to do it this way'*. That was the general attitude.

This example shows how the kind of work undertaken in The Room and its physical organisation sets up differences in work groups. Although different groups could be identified, there was a unifying way in which all controllers talked about their work, regardless of what kind of sector they worked.

Language and stories: Work as play and war

The language used in communication both conveys information and situates people in a social system (Resnick 1993). The language people use to describe their work experience reveals the ways in which their reflections shape their interpretation of their experience. It is necessary to first distinguish between the formalised language controllers use in the course of their work and the informal language they use to describe their work to others. The work activity of ATC involves use of a formalised and specialised language using standard phraseology to enable controllers to interact with pilots and other controllers. This standardised language allows each agent in the system to anticipate and respond to the requests and actions of the other agents. The clipped phraseologies used in the conduct of a flight are successful because as Weick (1987) suggests, they are anticipated and usually ratify expectations rather than inform. The analysis of the formal dialogue used in aviation work has been analysed at some length by others (Hutchins and Klausen 1996; Breul 2013) and is not the subject of this investigation. That research does, however, illustrate how individuals and groups tacitly share their understanding about work and how these tacit understandings guide practice. Informal language describing work situations and the myths and stereotypes that grow out of that language also provides evidence of what members of the group believe is important and what they have learned.

One of the important aspects of the way language is employed in ATC is what it reveals about the relationships between the agents in the system. Controllers talk of 'plugging in' to their 'slot' in The Room. Commencing on-the-job-training was described as 'getting a slot'. The term slot signifies the place the individual has, in relation to the 'state of play' of the system and the interdependency between agents. Perhaps because of its history and still close association with the military, the language used by air traffic controllers to describe their work and learning is often resonant of combat contexts. Policies and procedures governing assessment of learning, for example, are described as 'rules of engagement'. People 'fight' the system, controllers' talk of their work as 'going in to do battle' – against the air traffic that is encountered on the shift. A 'good' instructional relationship, for example, is based on an intimate knowing of how the trainee is 'living the battle'. From the perspective of the trainees, the experience of fighting the system can sometimes be particularly difficult.

I: You were saying you didn't want to come to work?
R: Oh yeah, that [phase of the training period] just felt like I was walking into the fourteenth round with five more to go with Muhammed Ali. Getting my head punched in rather than knocked out I suppose.

In this instance, the trainee is not fighting any one person in particular. They are fighting against themselves and their performance is being pitched against what 'the system' presents to them on their particular shift. Although this is a stressor for a trainee or someone inexperienced, it is the unpredictability of the traffic patterns when 'battling the system' and the problems that have to be solved that give many controllers their job satisfaction. Stories about working these periods of unpredictability (created by the different conditions every day and sometimes the periods of intensity and uncertainty) are sometimes turned into folklore (real and imagined). When shared as narratives of actual experience, they are described as 'war stories' and when used to project a particular type of controller, they represent archetypes of positive and negative performance ('gun controllers' and 'adrenalin junkies').

Shared narratives: War stories

War stories are the stories of a controller's experience where something dramatic happens, perhaps because of a controller's performance (or lack thereof), system deficiency or an unexpected event. War stories are passed informally between controllers and across Centres. Some of them are decades old, others are more recent. War stories are used to help illustrate both good and bad actions, right and wrong ways of operating. War stories have a dramaturgical quality. They often set up an 'us-against-other' struggle or battle of some sort and, therefore, are well characterised as 'war' stories. The 'other' may be another controller, the technology, the environment, oneself or a combination of these elements. For the purposes of this research, every controller who was asked about war stories talked about them in glowing terms. Not everyone could (or would) relay one within the confines of the audio-recorded interview. Given that war stories are often about the consequences of actions gone wrong, it may be that those who were unable to tell a war story were being reticent given that the interview was being recorded. The language used by controllers also reveals what individuals and groups value (or not) about their work activity.

Role of collectively held beliefs and values

In order to develop shared meaning, a group must work to create shared beliefs and values. A belief can be defined as a conviction or opinion, or an acceptance of something. Values are beliefs that we determine to be important (Wheelan 1994; Cameron and Quinn 2005). Values are underpinned by an evaluation about the worth, desirability or utility of something, or the qualities on which these depend. Values are ideals that arouse an emotional response, either for or against those evaluations.

Norms represent collective value judgements about how members should behave and what should be done in the group (Cameron and Quinn 2005). How do these elements of culture influence learning and performance and in what way? The following section will outline a number of collectively held beliefs found in ATC and then discuss the ways in which these beliefs become values in work performance when one's performance is evaluated in terms of the presence or absence of these attributes. This is not to suggest that all collective beliefs and values have been identified and are reported. The beliefs and values described here are best thought of as exemplars and will be used later in this book to illustrate their influence on learning and work activity. Three collectively held beliefs were identified and found in every Centre and within every sector and group. These were

> *The importance of ability*: A belief that ability is the foundation of exper-
> tise in ATC. Consistent with this emphasis on 'nature' rather than
> 'nurture' also were beliefs by some that controllers fit in a particular
> type of personality profile which consists of having certain attitudi-
> nal attributes such as arrogance, egoism and that these elements are
> necessary to do the job.
> *The importance of performance*: Performance is the way to demonstrate
> capability and self-worth. The emphasis on performance results in
> experience being regarded as the most valuable, and some would
> argue, only, way to learn ATC.
> *The importance of confidence*: That a necessary (but not sufficient) ele-
> ment of good controlling is confidence in the way one is undertaking
> the job.

It is contended that these collectively held beliefs and values influence both accredited and informal learning in the ATC workplace because they form part of the cultural context. They will now be outlined.

Importance of ability: Having 'The Right Stuff'

Tom Wolfe (1979), in his novel The Right Stuff, was one of the first people to popularise how 'natural' ability was valued in the aerospace industry. In his book, Wolfe described the culture within the US air force and the career paths that saw some pilots with The Right Stuff go on to become astronauts. 'At every level in one's progress up that staggeringly high pyramid, the world was once more divided into those men who had The Right Stuff to continue the climb and those who had to be left behind in the most obvious way' (Wolfe 1979, p. 30). That certain abilities are neces-sary to undertake a job such as controlling aircraft and that these capabili-ties can be screened in recruitment tests is well established (Hannan 1996;

Rantanen et al. 2006) and is not contested here. What is of interest here is the way some controllers and instructors believe that good controlling is obtained *only* by having The Right Stuff: that is, that the skills involved in ATC cannot be learned, and that good controllers are born, not made.

> I'm a great believer that flow controllers are born. You don't make 'em, they're born. It's an innate sense; they're very good at mental arithmetic. They're very good at looking spatially and they have very good sixth sense about whether it's going to work or not. And that's something you can't teach people and you don't get it out of a book.

As the controller above believes, good controllers are 'born and not made'. How one detected the presence of The Right Stuff also could not be easily identified by any of those interviewed. Identifying the elements that made up having The Right Stuff was a source of frustration also for many of the human resource developers and training department instructors interviewed. Training Department instructors spoke of how on-the-job instructors would complain that a particular trainee 'just doesn't have it'. According to this interviewee, when questioned about what 'it' was, the on-the-job training instructor concerned was not able to nominate anything in particular that the trainee was doing wrong. There is strong support in the learning theory literature for the claim that experts operate at a certain level of automaticity and, therefore, might not be able to consciously articulate what the 'it' is (Chi et al. 1988). In these cases, individuals who have been performing the task for a long time forget which maxims and rules (Benner et al. 1996) they are invoking in undertaking the work, making the skills learned opaque, even to themselves. Nevertheless, there are skills that can be learned and even skilful operators sometimes come upon novel situations, for which their existing abilities may provide no clues about how to proceed, requiring, therefore, engagement in learning.

Importance of performance

Confidence and ability are qualities embodied in good performance. Good performance in controlling traffic is what many controllers value in them and others. It enables those with The Right Stuff to show their value to others in The Room.

I: So they [controllers] then make those assumptions on those immediate facts as they're available, [but] they don't explicate them?

R: /ATC is very much the skilled craftsmen club and secret knowledge and all that sort of, not [right] down to secret handshakes, but as far as —you've [either] got it or you haven't. And you've got it by demonstrating it. So when I was a sector controller, I didn't know anything about ATC to an Arrivals controller — because I hadn't done Arrivals. So then I did Arrivals and I didn't know anything [to an Approach controller] because I hadn't done Approach. So then I went and did Perth Approach. And I don't know anything because the Sydney Approach control is much busier. So then I go [and do] Sydney control and 2 years later, I'm talking to a Perth controller ..., and he says *'oh you haven't done it* [Perth Approach] *for 2 years* [so], *you can't be current,* [so therefore] *you don't know anything about control.'* So it's very much that sort of — you demonstrate your acceptance [to the group], you know [their] tolerance of you, by the level of skill you do.

It is contended that what is regarded as 'good performance' is always relational. That is, it is always linked to how well one's performance is in relation to the performance of others. This is important in the sense that what constitutes 'good' performance is always set up, to some degree, by the group and not by external agents. Even objectivist accounts of competencies, for example, have those accounts developed by people with membership and credibility within the particular group or community of practice. In ATC, as in other occupations or activities, value (to the individual and to the work group) comes from how well the individual can perform the task, in relation to others. Demonstrating the qualities regarded as collectively important, such as doing the work fast and doing it well to the degree of complexity required of certain sites, are collectively held values of work in ATC. For example, a rating for a Sydney sector would be more highly regarded (by Sydney controllers) because the work in these sectors is more demanding since Sydney has a higher level of traffic per hour than other Centres. Likewise, having a rating on Approach would be regarded as more important (to Approach controllers) than having a rating in an Enroute airspace, because it is arguably more complex and demanding to operate.

Importance of confidence

A good controller is capable in their performance and confident in their belief in their own capacity to do the job. The following controller summarises the widely-held view of confidence and its relationship to the job.

I: Why is it that someone who didn't feel comfortable in and of themselves would still go through with having a check?
R: I think point to prove and ego, says a lot. I don't know if it's just ATCs in that we're all brought up or trained to be, [but] when I sit at that

console and I make a decision, I am 100% correct in what I am doing. I will reassess my decision, but every time I make a decision, I am correct in my judgement and I am never wrong.

In every interview, mention was made of the importance of confidence and its relationship to performance. Good controllers exercise good 'judgement' and have confidence in the decisions they have made. The issues of confidence and its relationship to self-efficacy and experience will be addressed more fully in Chapter 7.

In summary, three beliefs were widely held throughout ATC Centres about what it takes to be a good controller. These included ability (having The Right Stuff), the role of performance in demonstrating that ability and the importance of confidence in enabling the application of ability through performance.

It is contended that controllers require all three of these attributes in order to perform well. Beliefs also become values when they are applied to evaluate the worth of someone's work quality, that is, ability, capacity to perform and confidence are qualities on which good ATC work practice depends. These beliefs are illustrated as values when they become recurrent themes, shared through informal language and narrative in the process of collective remembering. Stories and language also translate into myths and legends which when used as resources to share understanding about someone's work activity, represent what is valued about work activity.

Myths and legends: Gun controllers and adrenalin junkies

According to the *Oxford Concise Dictionary* (2014), a myth is a traditional narrative, usually involving supernatural or imaginary persons, embodying popular ideas that are collectively held. Two mythological archetypes were commonly discussed in the interviews about ATC, and like many legends, the two archetypes identified represent the extremes of good and bad performance. In narrative terms, these archetypes represent the 'heroes' and the 'villains' of an organisation (Alasuutari 1995). A 'Gun Controller' is regarded as someone who demonstrates superb performance without even trying (such is their level of skill, ability and confidence). The Gun Controller has a limitless supply of energy, awareness and prescience. Unlike the controller depicted in the war story 'the breakdown of the labouring body' (Chapter 2), Gun Controllers have overcome the limitations of the body and are never exhausted.

I: What does a controller who's controlling with flair, or finesse, or a Gun Controller, — what do they do?

R: / A Gun Controller? Oh yeah, there's a few. They're just really good
controllers. They're on top of the situation all the time. They can handle a lot of traffic. They work well under pressure. There's probably in
Approach — there's a few. Not so many now. They were really good
controllers. I don't know if there are any Gun sector [Enroute] controllers. We don't see them. But the Guns were the Approach/departure
controllers and a few Gun Flows.

Gun Controllers do not even have to think about what they do, or to use
supports such as written calculations (as mere 'mortal' controllers may
have to). The interviewee went on to explain why the Flow Controller I
had sat in with earlier in the morning was not a Gun – because he used
a mathematical table he had prepared himself detailing calculations of
speed, distance from the airport and estimated time of arrival, which
aided him in his work activity. A Gun Controller, according to my infor-
mant, would not have needed such a resource, being able to do the calculations in split second timing with 100% accuracy, as and when needed
(just as quickly as the metaphorical 'top gun' would be able to draw his
weapon in a shoot-out). Gun Controllers, by definition, can handle anything, and do so with ease, flair and finesse. Although many controllers
believe in their existence, no Guns were identified in this investigation.
Not surprisingly, the people most likely to believe in Gun Controllers
were those at the apex of the cultural pyramid – Approach – at the apex
because their work most of all relies on the display of ability, performance
and confidence. The Enroute controllers questioned stated that they had
never seen a Gun Controller and even doubted their existence.

The negative construction of the Gun Controller is the Adrenalin
Junkie. An Adrenalin Junkie is someone who wants to be a top performer
(a Gun) but who doesn't have the ability to support the performance. The
Adrenalin Junkie gets a 'buzz' out of performing against 'the system'.
Perhaps the Adrenalin Junkie takes their position, their 'slot' (see earlier
in this chapter) in the state of play to an extreme. Adrenalin Junkies are
said to have forgotten that they are working with aeroplanes full of people
and see them as blips on a screen, similar to a video game. Controllers
gain satisfaction out of performing well, which involves moving traffic
expeditiously. The Adrenalin Junkie is said to have become addicted to
the adrenalin produced when performing under such demanding conditions at one's peak level. The following controller explains the dangers of
this approach and its relationship to both the real-time dynamic of the
work activity and the consequences for others.

> I don't think we have any other true adrenalin junk-
> ies. I think I might have been one once myself but
> not to this extent. I think the sort of people that we

> have, like myself, was we wouldn't say 'No'. We
> would keep taking aeroplanes on because, after a
> while, it is like a big sort of caramel lolly or some-
> thing – you can't break it up into small pieces, you
> just keep going and keep going and keep going. You
> can't bite it off or say *'Stop'.* You get to a certain level
> of busy-ness where you don't have the ability to say
> *'Hey. Stop!'*

In this case, the Adrenalin Junkie is likely to not notice his or her limitations have been reached and, in relation to the intensity and real-time flow
of the work, can get to a point where the work becomes too busy to split it
into smaller sectors. The importance of the value of knowing when one's
performance has reached its peak, and thus when one needs assistance
or to break the sector up into smaller components becomes an important
attribute in good work performance. It should be noted that although
people talked about Adrenalin Junkies, none were identified in this investigation, though like Gun Controllers, they were said to exist. Adrenalin
Junkies also represent the fine line that is needed between being confident and being cautious because ATC requires controllers not to spend
too much time thinking about the lives that are on board an aircraft, but
not to lose sight of that fact either.

> The thing is I think you will find that there are very
> few air traffic controllers who actually sit back there
> and think *'this is an aeroplane full of three hundred
> people'* and they are just blips on the screen. That
> is why ... because if they did feel that it was three
> hundred people, they probably couldn't do the job
> very well. Some people have gone so far to detach
> from what we are dealing with [and] they see it as
> a video game or a pinball game and they bunch
> them [aircraft] up close together and say *'this will be
> fine, this will be fine, this will be fine'* and [they are]
> cutting down the margins for error. That is one of
> the things we have to change. Okay, they are aero-
> planes with people on board. [But] we don't want
> them to go over-board and say *'Oh Christ, there are
> several hundred people on this aeroplane'.* At the same
> time we don't want them to say *'it's just a blip'.* It is
> somewhere in between. A happy medium.

In summary, the above discussion sets out a range of collectively held
beliefs about ATC work and illustrates the ways in which these beliefs

influence how people account for work behaviour and distinguish their behaviour from others as part of the cultural context of the work. In addition these beliefs are transformed into collectively held values about controlling work, as illustrated in the myths and legends and these represent archetypes of behaviour. It is contended that these archetypes are a form of projected occupational identity: the embodiment of an 'ideal type' – the Gun Controller – and the negative type – the Adrenalin Junkie.

Conclusion

This chapter discussed evidence of cultures in the ATC workplace. Within this work environment, a number of groups were identified based on values and interests external to the organisation (occupations, genders and other non-work interests) as well as interests and memberships within the organisation. Group membership is based on structural configurations as well as on values and beliefs that are collectively held. Group membership within the workplace involves identification with current work practice (such as a particular sector or Centre) and identification with a shared history of experience (such as those found within different divisions within the organisation). Collective values and beliefs are often upheld in narratives based on experience and these are sometimes translated into archetypes that can be evaluated for what they show is important within this particular work culture. The values and beliefs become group defining, with individuals deriving their sense of identity from collectively held beliefs and from group membership based on those beliefs. These are the contextual elements important to understand the *ghosts in the machine*, and this explication now enables the evaluation of their influence on the activity of learning.

chapter four

Rethinking learning

Over the past decade, there has been an increasing recognition of the importance of learning in the workplace and the need to develop appropriate theories and methodologies to understand it. It has been acknowledged, for example, that learning in the workplace occurs in a myriad of ways (Engestrom 2001; Billet 2002, 2016; Boud and Middleton 2003; Boud et al. 2006), including formal on-the-job training programmes, as well as through informal and incidental learning (Collin 2006). Learning through individual and collective experience is regarded as foundational (Boud and Middleton 2003). As a collaborative enterprise, workplace learning is practice bound and embedded in the everyday experiences of acting, negotiating and applying the problem solving skills which are part and parcel of the participatory process of working (Lave and Wenger 2005). Such learning is intertwined with the technical performance of work, its social networks being seen as a shared social practice (Schulz 2005; Collin 2006, 2008; Gherardi 2009).

Much of the literature on workplace learning has overlooked the ways in which organisational cultures mediate workplace learning, including instructor training. Collin (2008, p. 380) calls for research which explores the dynamic interactions between learners and instructors, thus providing a deeper understanding of phenomena which sustain or impede knowledge creation, distribution and reproduction. In addressing this issue, this book argues for a need to develop more appropriate cognitive–mediational approaches to instructor development rather than the receptive-accrual modes typically in use thereby both facilitating learner outcomes and enhancing organisational goals.

Psychological learning theories emphasised individual skill acquisition resulting in a change in cognition and behaviour as well as the development of expertise (Ericsson et al. 2006). Considerable discussion has been provided about the various kinds of learning (formal, informal, incidental) which occur in workplaces and whether or not even these conceptualisations help or hinder analysis (Billett 2002; Sawchuk 2008). In addition, Eraut (2004) has focused on the different kinds of knowledge employed in workplace settings and in particular the role of cultural knowledge. Further attention has been given to worker identity, and in particular the way newcomers learn and develop their workplace identities (Blaka and Filstad 2007; Timma 2007).

The concept of mediation is central to sociocultural theories of learning (Vygotsky 1978; Engestrom 2001). The way in which mediation is used in this chapter is drawn from the work of Vygotsky (1978) where an acting subject engaging in object-oriented activity is able to learn as a result of mediating tools (e.g. cultural meanings, support from knowledgeable others) which in turn leads to development and change.

In this respect, mediation comes into play in a range of ways, such as in cultural ways of knowing (Billett 2016; Resnick 1993) as well as through the use of physical artefacts (as in the case of distributed cognition – Hutchins and Klausen 1996).

In learning theory terms, the notion of constructivism suggests that all experience and learning are filtered by what the individual currently understands and believes. Constructivism has two forms: cognitive constructivism, which focuses on an individual's internal schemas and mental models for making sense of the world; and social constructivism, which emphasises the role of the social context in shaping what is learned (Appleton 1996). In this book, the idea of cognitive–mediation is used to represent the ways in which what an individual learns from an experience is filtered by the interpretation of that experience and shaped by particular collectively held beliefs and values found in organisational culture.

From this point of view, acting subjects (e.g. learners, instructors), when engaged in their workplace activity (learning/instruction) will use strategies that are mediated by collectively-held values and beliefs and other artefacts of organisational culture, such as stereotypes and norms of practice (Wertsch 1998; Engestrom 2004).

Learning theories based on the concept of mediation (such as constructivism), challenge the notion of learning through typical means of reception and accrual instruction because individual and cultural beliefs will enable and constrain what is observed, noticed and thus remembered (Resnick 1993).

What is important about these contributions is that they illustrate the ways in which learning in the context of work activity is a product of the interactions we have with others. Extrapolating from these theories, ideas or concepts are essentially cognitive tools with which we think and reason. It is through sustained interaction that individuals come to share common ways of thinking and expressing ideas. Such communities who share ways of thinking and communicating are sometimes called 'discourse communities' or 'communities of practice' (Gherardi and Nicolini 2002; Blaka and Filstad 2007). From this view, learning is a matter of becoming socialised into a certain way of thinking and learning how to use certain concepts, skills and procedures in discourse with others. In a workplace context, for example, the trainee learns not just from explicit instruction, but also through observation of and interaction with more knowledgeable members of the group's culture as affordances (Ellinger and Cseh 2007). However, focus should not be limited to learners alone. Instructors too develop appropriate ways of instructing from others.

Formal and informal learning

Learning is easily visible when it is formally organised, where it often leads to the accreditation of workplace skills, and yet it also occurs in a range of non-formal ways that are not so easily visible but are just as important. In the air traffic control environment, both formal and informal learning are increasingly important. Both formal and informal learning are influenced by the workplace context, defined here as the organisational structures and cultures that are inherent components of all workplace environments.

Defining learning in the workplace

A broad definition of learning is needed, therefore, to examine the influence of contexts on: (1) individual learning occurring in skill acquisition and (2) the activity of learning embedded within work activity in the workplace. In this context, *learning* is defined as the process involving the transformation of experience through reflection, conceptualisation and experimentation which leads to an increased capacity in individuals, groups and organisational systems, to act in the environment (after Kolb 1984). As such, learning has the following features:

- It is actively constructed by individuals and groups, based on their interpretation of their experiences in relation to their histories and the relationships between them and the social world (Putnam and Borko 1997).
- It is both a process and a product of the interactions we have with others (Resnick, Levine and Teasley 1993).
- It is embedded in contexts which are socially and historically constituted and are linked to broader societal structures and systems (Engestrom 2001, 2004).
- It is often mediated by artefacts (i.e. by technologies such as computers, tools and resources). As such, learning is distributed across individuals, other persons and the physical environment (Salomon 1993).
- It is sometimes contested. Learning, particularly in organisations, involves people whose social locations, interests, reasons and subjective possibilities are different, and who improvise struggles in situated ways with each other (Holland and Lave 2001).
- It has an affective dimension, because it involves the active engagement of the self – the learner. The affective dimension of learning

often arises from discrepancies between knowing and experience, which in turn generates dissonance (Brookfield et al. 2006).

- It is a matter of degree. It can be reproductive (of past practices, beliefs or attitudes), adaptive (i.e. incremental adjustment) or it can be transformative (leading to the creation of new knowledge and a reconceptualisation of understanding and changed action) (Schon 1991; Argryis 2004).

Abstracting a learning framework

The purpose of this chapter is to develop a framework that can then be used throughout the rest of the book to analyse and discuss the ways in which contexts are implicated in processes of learning in the air traffic control workplace. However, according to Latour (1987), problems arise when abstractions are cut off from the elements that tie them together. They 'float like flying saucers' (p. 242) above the material world that was once their basis. So, duly warned and proceeding with caution, a framework will be presented as a means of linking organisational and learning theory and to set up the structure for the findings in chapters that follow.

The experiential theory of learning by David Kolb has been chosen as the analytical framework because it highlights the importance of experience and emphasises learning as a process and a cycle – a notion central to the goal of continuous learning. It will be argued that Kolb's theory, with modification, provides a useful framework for integrating many of the elements relevant from both psychological and sociocultural theories of learning. Moreover, it also provides a useful means of then linking these elements of learning to structural and cultural aspects of workplace contexts. Before identifying the modifications needed to Kolb's theory, its utility and weaknesses will first be outlined.

Kolb's work draws on Dewey (1933, 1938), Lewin (1951) and Piaget (1960) in contending that learning occurs as a result of a resolution of a contradiction or conflict between opposing ways of dealing with the world: between reflection and action and between doing and thinking. Underlying these processes of learning about the world is the notion of apprehending (or grasping understanding) and comprehending (understanding and moving on). For Kolb (1984), there are four elements involved in the learning process (experience, reflection, conceptualising and experimenting) and each of these must occur for effective learning.

As a framework for accounting for learning, Kolb's model has a number of strengths as well as weaknesses. An important strength is that Kolb does not make the mistake of equating learning with the acquisition of knowledge. For Kolb, learning must be evident in what people do. Second, in organisations, problems are both the stimulus and the medium for learning. In this respect, Kolb's model is directly applicable to conceptualising how people in organisations may experience problems and learn from them. Third, his model draws on explicit phases that are necessary for effective learning to occur and thus provides a useful framework for explicating processes of learning.

However, Kolb's model also has a number of weaknesses. Although Kolb (1984) discussed the dialectical movement between the various processes involved in learning, Kolb's work has been popularised as a set of four 'stages' involved in the learning cycle, and these are presented almost as if they are discrete entities that occur in a linear cycle. It is contended that the very nature of a dialectical process means that engagement may move back and forth between processes such as reflection and experience (or any other parts of the cycle) before progress in learning is achieved. Thus, these four elements of the learning cycle identified by Kolb are very difficult to tease out as distinct phases and may be fused within a particular activity. However, for the purposes of expediency and clarity the four elements are presented individually although this is not to suggest that they are discrete or occur linearly.

The second weakness is that Kolb extends his experiential theory learning to develop a series of static learning 'styles' that individuals are said to possess. Indeed, most of the usage made of Kolb's work is based on this notion of learning 'types' or styles (Dyrud 1997; Knight et al. 1997). The concept of 'learning style' is not relevant to this discussion. It is argued that it is neither helpful nor desirable to pursue such a notion when looking for ways to enable individuals, groups and organisations to engage in continuous learning in changing environments. The focus here is on investigating the influence of contexts on enabling or constraining processes involved in learning. Therefore, the concept drawn from Kolb's theory relevant to this discussion is the notion of learning as a cyclical process which involves the four stages mentioned above. However, these stages too need to be modified and this raises the third weakness in Kolb's model. It is argued that the terms Kolb uses to frame learning (*concrete* experience, reflective *observation, abstract* conceptualisation, *active* experimentation) are unduly limiting. It is contended that each of these aspects of the learning process can involve dimensions other than ones that are concrete (in terms of experience), or based on observation (in terms of reflection). Varying levels of abstraction in terms of conceptualisation and experimentation might involve the teasing out of 'what if' statements rather than direct action. Therefore, for the purposes of this analysis and discussion, the following definitions of the four elements are offered:

- Experience occurs as part of being-in the world and may be registered through what we see, think, hear and feel as we interact with our environment. It may occur directly or may be experienced indirectly through others (e.g. vicarious experience).

- Reflection is the process involved when attending to, noticing, recalling elements that are significant in the experience and can be either passive or active.
- Conceptualisation is the process of thinking about, making sense of, interpreting and comprehending those reflections and experiences.
- Experimentation is the activity associated with developing choices and envisioning new ways of acting that may have occurred in the past, be occurring in the present or likely to occur in the future.

The final weakness of Kolb's theory is that, like other psychological approaches, it ignores social context and thus emphasises learning as an internal individualistic process. The main purpose here is to demonstrate the embeddedness of learning and context. As a structure, Kolb's model, therefore, needs to be able to illustrate the influence of contexts on learning in the workplace.

From a sociocultural perspective, the social environment – the work group and other communities of practice – also influences learning (social cognition). Learning is mediated by elements found within the immediate situation (situated cognition). The tools and physical resources people have available to them (distributed cognition) would also structure their learning, as would histories of previous experiences of people and their structural roles within activity systems (activity theory).

The following table summarises the main elements of the framework that will be discussed in detail in the following chapters. Table 4.1 maps the linkage of contexts with the various processes identified as important in learning.

Table 4.1 expands on the four processes important in Kolb's theory of learning and links them with appropriate contextual features important in air traffic controllers' work structure and culture. The table lists the attributes identified as important within each of the four learning processes. The next column then summarises the structures that were identified in the study as important in influencing each learning process. The cultural analogues that are influential are listed in the next column. In terms of experience, it will be shown in subsequent chapters how these cultural elements are marked, remembered and accounted for through reflection and conceptualisation on experience, leading to experimentation.

It is important to point out that the table *summarises* structural and cultural elements regarded as important within each of the processes of learning. Therefore, items listed in the framework are not exhaustive or mutually exclusive. For example, as a structural element, formalisation is implicated in the data as influencing the process of reflection, as well as experimentation. Similarly, as a form of work organisation, teamwork both enables and constrains opportunities for reflection (indicated by + and −). Likewise, both reflection and conceptualisation are occurring

Table 4.1 The linkage between elements of context and processes involved in learning

Learning processes		Attributes	Structural analogues	Cultural analogues
Experiencing	Being	 • Corporeally (body) • Cognitively (mind) • Affectively (self) • Socially (environment) COP	Physical work organisation Dimensions of work experience: temporally, complexity, affectively, socially Differentiation and integration of job tasks and roles	⇐ Marked, remembered, accounted for and generated through ⇓
Reflecting	Observing	 • Narrating/remembering; noticing; pattern seeking; labelling	Work organisation Integration structures • Job roles (+) • Teams (+ −) Formalisation	Collective remembering (e.g. war stories) (+) Shared meaning (evident in informal language used by groups/symbols
Conceptualising	Thinking about, sense making	 • Schemas used to interpret, explain, account for • (re)framing • Pattern generating	Integrative structures (e.g. boundary spanning functions) Job Roles	Collective remembering Collective schemas • Shared beliefs and values (stereotypes) • Shared norms
Experimenting	Acting	 • Envisaging/enacting • Developing choices validating/testing • (re)evaluating • Adapting/innovating • Obtaining feedback	Integrative structures • Teams • Boundary spanning Centralisation Formalisation	Norms of practice (+ −) Collective remembering (+) Collective schemas (e.g. stereotypes) (−)

Note: +: Enabling; −: Constraining.

when narration of stories ('war stories') are used as a resource for learning. Although the cycle of learning proposed by Kolb could start anywhere, a useful starting point is with experience, since, it is contended, experience is central to learning (Boud and Miller 1996; Brookfield et al. 2006).

Experience as a process of learning

According to Boud and Walker (1993), 'experience is created in the trans-action between the learner and the milieu in which he or she operates – it is relational' (p. 11). Hence, it is contended, experience is mediated by organising devices (structures) and interpretive processes (cultures). Experiencing is always grounded in some context and is, therefore, influenced by artefacts. In the workplace context, for example, experiences are structured by the artefacts used in work organisation, such as the physical resources, policies governing activity which structure workplace experience and opportunities for experience. However, experience also involves perception, implies consciousness and always comes with meaning (Boud and Walker 1993). Given that interpretation of meaning is the foundation for culture, culture is thus always embedded in the interpretation of experience. The structuring and interpreting of experience will influence opportunities for learning in certain ways because structures will make certain opportunities available and not others and cultures will focus attention on particular interpretations of the experience and not others.

In the research, the structuring of ATC work was observed to influence experience in particular ways. Drawing on the components of how we experience the world (cognitively, corporeally, affectively and environmentally) it will be argued in Chapter 5 that particular aspects of ATC work structure emphasise various aspects of experiencing. For example, the temporal dimension of experience is emphasised in the 'real time' dynamic of the job. Complexity is evident in the nature of job tasks that require higher order thinking and problem solving. An affective dimension is also evident in, for example, the importance given to individual and collective decision-making as an expression of self and group. Finally, a social dimension is evident in the interdependence of job tasks. It will be argued in Chapter 5 that these dimensions of experience are characteristic of work generally, but especially important in High-3 work in particular. Work may be intense on the body; easy or demanding on the mind; an expression of, or alienation of, the self and it always occurs in a social environment, although some forms of work organisation provide more emphasis of the 'social' than others. It is argued that the way work is organised shapes opportunities for experience in certain ways and that these organisational structures influence the transformation of experience into learning by enabling or constraining the transition to the next moment in the cycle – reflection.

Importance of reflection for learning in the workplace

In learning theory terms, reflection is an essential ingredient for learning (Schon 1991; Brookfield et al. 2006). Reflection involves attending to the salient features of an experience, marking and noting those features through labelling and pattern seeking. It consists of those moments in which individuals engage in to recapture, observe and notice and begin to make sense of their experience, 'to work with their experience to turn it into learning' (Mason 1993, p. 9). It will be contended in Chapter 6 that there is a close link between individual and collective processes of both reflection and the next phase of the learning cycle, conceptualisation (or making sense of reflections on experience).

In workplace contexts, cultures are based on the salient features of experience which are noted, collectively remembered and attended to through narration and through the common language and symbols that groups use to share and express their experiences with others. Individually, the degree to which reflection is engaged in will depend on the motivational levels of the person to make sense of the experience and their interpretation will be influenced by their sense of self-efficacy. In the research, reflection was observed to occur culturally, as a part of shared remembering evident in the telling of war stories. War stories, as well as the informal language controllers use to describe their experiences, are a collective process of reflection because they draw attention to what is worth noticing, and thus worth learning, and what is not. In addition to these cultural processes, organisational structures also enable and constrain opportunities for reflection.

In terms of structure, capacities for reflection are variously designed into job roles. The form of work organisation characterised by Taylorism, for example, was predicated on the basis of removing opportunities for reflection (and conceptualisation) from certain job roles. In the research, the way the work was structured both enabled and constrained opportunities for reflection. Reflection through observation, for example, was enabled through the visibility of work designed into the workspace and through the organisation of job tasks and roles (e.g. on-the-job instruction). Organisational structures also constrained opportunities for reflection on experience, however, through the organisation of work activity where the experience of the temporal dimension of work (e.g. its intensity and immediacy) limited opportunities for stepping back and reflecting on what was happening. In these contexts, cultural processes of reflection, found in shared remembering through narration, become particularly important as a compensatory mechanism for the ways in which structures limit this process of reflection. Structures and cultures also enable and constrain the next process in the learning cycle, conceptualisation.

Role of conceptualisation for learning in the workplace

The next phase in the cycle, conceptualising, is the process of making sense of what has occurred, to interpret reflections on experience and generalise these interpretations to new settings. Many learning theorists find it difficult to tease out reflecting from conceptualising, since the process of making meaning relies on the interconnection of noticing, interpreting and making sense (Argyris 2004; Augoustinos et al. 2014). Nevertheless, the processes of reflection and conceptualisation in learning have been separated although it is acknowledged that in the everyday world these elements are closely intertwined. The conceptualising process begins with the elements of reflection (observing, noticing and labelling which triggers remembrance). Conceptualising is a practice whereby the meaning from the experience is generated into concepts or ideas that can apply to situations beyond an explanation of the immediate experience. As such, conceptualising also has been called theorising. Mason (1993) commented that the term *theorising* is based on the Greek root *theoria*, meaning *a way of seeing* and *abstracting*. One of the meanings of abstracting is the search for or distillation of essence or structure. Mason's comments demonstrate also the close linkage between the processes of conceptualisation and experimentation – the next element in the learning cycle, since noticing and interpreting leads to the development of alternatives that can be used in the future. Conceptualising or generalising from one experience to another involves identifying patterns in experience found through reflecting and generating ideas about those patterns in other events. Organisation and categorisation of perceptions enable comprehension and interpretation of the social world. In theories of 'reflective learning' or 'reflection-in-action' (Schon 1983, 1987, 1991; Argyris 2004), the elements of reflection and conceptualisation are evident but not separate. For these theorists, the processes of reflecting and conceptualising together are called 'reframing'. That is, when an initial perception is transformed into a new understanding or frame.

Sometimes conceptualising or reframing is constrained, however, by what is observed or noticed, a feature that is important in social cognition and constructivism. For Resnick (1993), interpretation of experience is based on schemas that both enable and constrain individuals' processes of sense-making. A schema provides an interpretive framework that allows reasoning to proceed (Resnick 1993).

As an interpretive framework, a schema is often based on past history, sets up expectations about what will be important, and therefore, will help guide what we attend to, what is perceived, what will be remembered and what will be inferred.

What this illustrates is that schemas are not purely individual constructions but are heavily influenced by the kinds of beliefs and reasoning schemas available in the individuals' surrounding culture (Resnick 1993). Individual and collective schemas are thus obviously mediated by cultural contexts, since organisational cultures are based on collectively held beliefs and values (where schemas or shared mental models are embedded) and these in turn generate norms of practice (Augoustinos et al. 2014). It will be argued in Chapter 7 that the contextual elements of organisational culture (such as collectively held beliefs, values and norms) are resources used in the conceptualisation process because they influence perceptual selectivity and the development of individual and collective schemas about work practice. In addition to influencing events, collectively held beliefs and values also will be reproduced by individuals and groups in the kinds of continuous learning strategies used in work activity. Collective beliefs and values, therefore, will lead to some things being noticed – and indeed emphasised – rather than others. Thus, schemas enable and constrain both individual and collective opportunities for learning.

In addition to culture, conceptualisation is influenced also by organisational structures when the activities of explaining, accounting for and pattern generating are built into job tasks and roles. Clearly in complex organisations, characterised by the need for high reliability, where the demands of work extend beyond any individual's capacity, the responsibility for task completion rests not just with the individual, but is also placed within the group. This is, particularly, in work environments designed with high levels of interdependence in work activity. Indeed, within the human factors literature, 'shared mental models' has been a recent focus of research, which involves investigating performance in High-3 organisations (Hoc and Carlier 2002; Hollnagel and Woods 2005; Soraji et al. 2012). Such research has pointed to the ways in which unstated interpretations (mental models or schemas) become assumptions upon which action is based, sometimes with devastating results. Nevertheless, opportunities for experimentation are evident in a range of ways.

Experimenting in the workplace

The fourth phase in the learning cycle is experimentation. According to Kolb (1984), learning is limited if an individual formulates concepts to generalise to other settings, but fails to test their validity. It is contended that testing the validity of conceptualisations, based on reflections on experience, can be done through evaluating past experience and envisaging new alternatives to be put into action immediately or sometime in the future.

Envisaging new alternatives may occur also in thinking about past actions (reframing). In this case an expansion of the range of choices available might be made though they may or may not be acted upon in the future. Therefore, for the purposes of this discussion, the term 'experimenting' has been emphasised because it can involve reframing actions that occurred in the past, action to be taken in the present and also it can mean developing choices to put into action in the future. Experimenting then, refers to developing choices and envisaging new ways of acting. These choices and alternatives are tested out mentally and/or practically through developing alternative plans of action for the future and acting on those alternatives when appropriate.

Expanding the range of choices and alternatives available is mediated by conceptualisation, which, as previously discussed, involves generalising concepts to new settings. Argyris (2004) proposed the idea of 'single loop' and 'double-loop learning' as an indicator of the ways in which developing choices for future action was mediated by conceptualisation. Individuals and groups involved in single loop learning have limited alternatives and possibilities for action available to them because they operate without testing the basis of their understandings and assumptions (based on their beliefs and values) about the problem or situation. When organisational members have the awareness and ability to recognise their underlying assumptions and to test their validity, and modify their thinking, double loop learning occurs.

It will be argued in Chapter 8 that organisational contexts enable and constrain opportunities for experimentation for both individuals and groups. Structures such as teamwork, for example, increase possibilities for experimentation because they enable a shared continuity of experience to occur across team members and thereby enable the experience to be used as a resource for inquiry to generate increased possibilities for action. Similarly, the degree of formalisation within a workplace may limit the capacity for individual experimentation though policies and rules formalising work activity may also embed behaviours aimed at generating alternatives into job tasks and roles. Organisational culture influences individual and group opportunities for experimentation to the degree that such practices are enabled and constrained by collective norms of practice, shared conceptual schemas that account for how the world works as well as shared capacities for collective remembering.

Conclusion

The chapter has outlined the components of the framework – four specific learning processes identified by Kolb – that will be used to investigate and analyse the data in subsequent chapters to determine the influence of contexts and how they enhance or inhibit learning in the workplace.

The chapters ahead will show that organisation contexts influence opportunities for learning in the workplace when they enable or constrain:

- Individual and collective involvement in activities associated with each of the four processes involved in learning

- Transition between the processes and integration of them
- The transfer of insights from engagement in these learning processes within and between individuals, groups and organisational systems

These are the elements that must be attended to if continuous learning is to be a goal of the workplace.

chapter five

Experiencing ATC work

Embedded nature of experience in learning: A cautionary note

It is contended that the very nature of *experiencing* is likely to have embedded within it the processes of reflection, conceptualisation and experimentation. However, for the purposes of this book, even though it is acknowledged that these processes are closely interlinked. For the sake of clarity each of these phases will be examined separately, something akin to a snapshot. The challenge, then, is to identify the influence of contexts on enhancing and inhibiting the transformation of experience into learning. That is, to identify the ways in which structures and cultures enable or constrain transitions between experience, reflection, conceptualisation and experimenting. The first step, however, is to identify the influence of context on the various elements of experience within the ATC workplace.

Experiencing ATC work

In the ATC workplace, organisational structures shape experience across four dimensions. These dimensions of experience are listed in the first column of Table 5.1. Analysis of the data found that work is structured so that the temporal nature of experience is emphasised. Temporally demanding work occurs in a dynamic 'real-time' environment; it involves periods of intensity, working within short time frames and it requires skills in concentration and immediacy (involving 'living in the now'). This dimension of experience is often registered in the body. Second, the work is structured so that it is also experienced complexly. Complex work requires the coordination of multiple tasks that in turn require higher order thinking and in combination with the temporality of work, practice is needed to build up awareness and understanding of the various permutations of problems and problem solutions that successful task completion may require. This dimension of experience uses the mind and thus cognition. Third, work is experienced affectively. That is, the act of working involves the self and is an expression of the self to others. Individuals both contribute to effect (and affect) the work undertaken and the work, in turn, affects the individual. From work one gains a sense of self-esteem and identity. In High-3 work, one works intensely

Table 5.1 Work experience and its impact on ATC workplace learning

Dimensions of experience	Contextual features/structures indicated	Evident in aspects of work performance	Learning work[a]
Temporal: *real time, short term; intense* (Corporeal – body)	Way work is organised; Nature of work activity	Use of body (body-clock monitoring)	Body-clock development; Anticipation
Complex: *work strategies in decision-making* (Cognitive – mind)	Formalisation (rules and procedures, for example, policies to reduce complexity)	Level of job autonomy and decision-making; Using artefacts to aid in problem-solving	Problem recognition; Prioritising; Problem-solving development of multiple strategies
Affective *presentation of the self, evaluation of performance of self* (Psychological – self and identity)	Nature of work – high reliability; Physical work organisation – visibility	The embeddedness of performance with affect: confidence	Confidence (impression management); Self-awareness; Independence; Learning limitations
Social: *interdependence, visibility, system interdependence* (Interaction with environment)	Physical work organisation (spatio-locations of sectors); Integration of job tasks; Teams	Strategies to monitor interdependence; The requirement of authentic performance	Developing the third ear; Developing situational awareness of others; Learning who to trust

a What instructors look for in trainees during on-the-job instruction.

at a complex activity that includes risk and once commenced, cannot be easily stopped. The risks taken in undertaking the work are publicly available for others to observe and involve a sense of one's own self-efficacy (Bandura 1997, 2001). This dimension of experience involves the psychological self (Fiske and Taylor 2013). Finally, work is experienced socially. That is, work is shared and is organised such that each individual is interdependent with others who form groups which in turn are interdependent with other groups.

Table 5.1 summarises from the data the structural aspects of work organisation that give rise to the various dimensions discussed above and how these become evident in work performance. The table also summarises what instructors look for when monitoring and assessing development in accredited learning. These dimensions of experience interact with one another in the course of the work and these interactions are summarised in Table 5.2.

Table 5.2 shows, in italics, the aspects of experience that arise from the physical structuring of work organisation (e.g. intensity) that were discussed in Table 5.1. Table 5.2 shows the attributes controllers must learn to develop in accredited learning and use when engaging in informal learning on the job. The attributes represent the different combinations of the dimensions of experience. Reading diagonally across the table, from left to right (identified in the table by the bold boxes), are the attributes that are important within each dimension of experience. The temporality of work, for example, requires responsiveness as a result of intensity; its complexity leads to variability; its affectivity leads to physiological and psychological arousal and the sociality of work leads to systemic interdependence of the various actors involved in the work system. The table also lists those aspects that are identified as the attributes that good controllers must develop and demonstrate (e.g. body clock development that enables the controller to establish accuracy with timing) and these aspects will be discussed throughout the chapter. The table also lists the various combinations of dimensions. The attribute of perseverance, for example, is needed as a result of the work being experienced as both temporally, complexly and affectively (represented in the table by the shaded area). The attributes referred to later in this chapter as 'interdependent sentience' are needed as a means of experiencing work temporally, complexly, affectively and socially. These attributes are presented in this way because, as was discussed in Chapter 1, Introduction, as indicated by the metaphor *Ghosts in the machine*, the ways of experiencing the world are not presented as dualisms (mind/body; self/environment; individual/group) but are recognised as inherently linked together and inseparable. This assumption is supported by sociocultural theories of learning (Suchman 1987, 1996; Lave 1996) and is the basis also for assuming that contexts are vitally important in workplace learning. Nevertheless, for the purposes of

Table 5.2 Elements of experience and attributes displayed in good ATC work

	Temporal	Complex	Affective	Social
Temporal	*Responsiveness* *Intensity* Accurate timing (body clock) Concentration Vigilance (scanning) Awareness and monitoring	Perseverance		
Complex	Situation awareness Pattern recognition Longevity	Variability Visualisation Judgement Holistic understanding	Interdependent sentience: • Referential anchoring • Work synchronicity • Collective memory	
Affective	Awareness of own strengths and limitations confidence Self-presentation	Authority Self-efficacy Impression management	Physiological and psychological arousal Involvement of self	
Social	Strategies to monitor flow in relationship with others third ear	Interdependent understanding awareness of consequences for others	Individual and social identity	Systemic interdependence Visibility

this book these dimensions will be discussed individually with reference to the various combinations when appropriate.

Temporality of ATC work

ATC work occurs in a dynamic 'real-time' environment. That is, the work cannot be stopped; the problems confronting the controller must be resolved. Time and time pressure are key features of High-3 work. This time pressure can sometimes lead to a sense of urgency; though urgency may also be a routine part of the work (recall Chapter 2 where the time an aircraft may be in a sector was limited in some cases to three minutes). This is one of the elements of work that controllers use to distinguish the uniqueness of their work from the work of others, as the following controller explains.

> If you've got three or four intercom buttons flashing that's a really good sign that you're getting behind. Because you should be able to punch those and get rid of them really quickly and not get to the stage that you're that overloaded. If you've got four buttons flashing, that could be two minutes of work and if you're two minutes behind, you're a long way behind.

This quote illustrates both the time pressures involved and the short term nature of task completion. These two attributes give the work experience its intensity.

Using artefacts to monitor the temporal dimension of the work

Theorists involved with distributed cognition emphasise the ways in which objects and displays function to support collaborative mental work (Pea 1993; Hutchins and Klausen 1996). A number of strategies are used to monitor the temporal dimension of the work by individuals and groups and these strategies involve a range of artefacts. As the quotes earlier have illustrated, controllers use the lights flashing on their consoles as indicators of their performance in relation to the ebb and flow of the work. These cues allow controllers to monitor the work and to anticipate when extra help may be needed before the situation becomes too busy.

In ATC, the temporal experience of work intersects with the social experience (Table 5.2) as ATCs use shared spaces and shared objects to monitor the flow of traffic. Shared objects and displays facilitate the process of what Resnick (1993, p. 10) calls 'referential anchoring'. In ATC work, the visibility of workload on other sectors enables controllers to referentially anchor their work in relation to what can be anticipated. For example, hearing the build-up of traffic on a neighbouring console provides a common referent that can be used by controllers to monitor what work is building up that is likely to be coming their way. Because the information about the activity of work is publicly available, team leaders and colleagues can also anticipate when a controller working at the console is likely to need help.

Using the body as an artefact to aid performance

Joas (1996) contends that most theories of action overlook the role of the body and are based on an implicit assumption that 'the body is the factual basis of action but pay no attention to it, as if in a fit of theoretical prudishness' (p. 167). Joas (1996) goes on to argue that the role of the corporeal body in action deserves greater attention. A controller uses his or her own body to monitor performance and to determine whether he or she is getting overloaded, as the following controller explains.

I: How do you know you're getting close to the edge?

R: Personally? I start to perspire basically, that will be one of the major indications - I'll start feeling hot. I think you'll find a number of people will say the same thing. I notice a couple of the girls - if you walk in and they're red in the face you know they're starting to push it – that's a physical sign. Another physical sign is you'll notice that you're not saying call signs quite as well because you're trying to say them too quickly. Little things like that – that should be ringing bells in your head. Either that it's time to get out or it's time to get someone else in.

This quote highlights how the controller uses their body to monitor their workload. It is contended that the use of the body in this way is a mechanism to manage work in a temporally based and intensive environment. Controllers also use their bodies to aid performance. In the interviews, controllers would talk about developing their own 'body clocks' – to know when an aircraft is due to call, and when that moment has passed (and thus when remedial search and rescue action might be needed). In accredited on-the-job training, body-clock development represents an indicator of the controller's emerging understanding of the temporal flow of the work (Tables 5.1 and 5.2), as the controller begins to build up the skills required for expertise and accompanying automaticity. Instructors look for trainee anticipation of when something should happen as evidence of progress in body-clock development. Being surprised by a call indicates a lack of awareness of the air traffic situation and is evidence of poor body-clock development. In the past, instructors talked of 'ruining' time clocks so that the trainee had to build up their own body clock.

Controllers use also other parts of their bodies as aids in performance. The intersection between the temporal and affective dimension of experience (Table 5.2) requires controllers to know when they are getting to the edge of their limitations and to use the voice to display confidence. Controllers discipline their voice, for example, to sound calm and confident and thus to project authority and control into a situation. This is one of the first things a trainee must learn. Sounding confident is important because it has an impact on the operation of the entire system. In the following excerpt, the respondent has been called on by a colleague to assist a trainee to overcome her lack of confidence. In doing so, he explains the impact a lack of confidence will have on the controller's workload.

I: So he (the instructor) came to you for your assistance (with the trainee)?
R: We told her that it doesn't matter if that isn't exactly what you think is the perfect thing to do. If it is safe, be confident in it and do it, because if you hesitate, then things will get worse. You've got to be confident in your memory, what you remember, how you approach things. You've got to achieve the pilot's confidence straight away. He doesn't give you a second chance. If you do something and he feels..., like.... He's... putting his life in your hands, and if your confidence isn't there, or he doesn't feel it's there, then he's going to be nervous. And that's going to make things more difficult, because he's going to want to know more information, then you're going to get further behind, and it steam-rolls itself. So if you're not confident, those around you and those you speak to will hear it and you'll undermine yourself, and your confidence gets worse. It's a difficult time for somebody who's been away, or someone who's had an incident, or who's newly rated, to sound confident. Especially for some of the younger trainees, because they sound so young.

Controllers must be confident in their own decisions so that they can move on to the next problem and they must communicate confidence to others. However, with the cultivation of confidence must also come a cultivation of doubt, as controllers need to maintain vigilance against what they call a 'fat, dumb and happy' attitude of complacency. This is a balance between overconfidence and cautiousness. Controllers, therefore, need to be both confident in their decisions as well as constantly checking and scanning the situation to notice unanticipated disturbances. This involves having both temporal and cognitive awareness of the situation (Table 5.2) as controllers require skills in concentration and vigilance. Overconfident controllers can make a mistake because something is not checked and underconfident controllers can make a mistake because they undermine their performance. Lack of personal and interpersonal confidence will create problems because it will create extra work and as a consequence lead to performance decline.

Complexity of ATC work

In ATC work, decisions are made, changed and remade as information becomes available. The work requires combinations of anticipation, planning and decision-making as various problem permutations become available. The nature of service work is such that a task or situation will rarely have the same combinations of features a second time as is evident in the following quotation.

R: You might come to work one Friday, and everybody gets away on time. They all cruise along, weather's fine, no turbulence. You go home and think *'did I actually go to work today?*

Nothing happened!' And the next Friday, you expect the same scenario but two are late, one's early, and one's on time. And they're all in a clump. And the weather's bad, and you've got four aircraft and you have to sort it out. You can't simulate the way it happens in The Room.

Controllers operating in this work environment must be open to variability and act responsively and flexibly (Table 5.2) in working through the permutations that occur as part of the 'routine trouble' (Suchman 1996) encountered within the work activity. They need to build a holistic understanding of the air traffic pattern, relating the individual problems back to the whole. Development of these skills is supported by psychological theories of expertise.

Gaining 'the picture': Using artefacts to aid in visualisation

Research into ATC expertise (e.g. Redding et al. 1992; Taylor et al. 2005) suggests that controllers maintain a mental air traffic three-dimensional 'picture' which is a visual representation of the air traffic pattern at any one time. Controllers check the accuracy of their visualisation of the air traffic flow (particularly in procedural non-radar sectors) by anticipating when aircraft travelling in opposite directions will sight and pass another. To check this they will request the pilot to notify the controller at the exact time this occurred (having already prepared an estimate of when the event will occur). The use of imagery has been investigated also for its role in maintaining situation awareness (e.g. Endsley 1994; Issac 1994). However, it is contended in this book that controllers also extend their use of visualisation to incorporate their manipulation of the tools they use in undertaking the work. They do this by beginning to represent parts of the airspace with parts of the console. This enables them to maintain a visualisation of the airspace and to prioritise their work, as the following instructor explains.

I: So you're working out where it [the aircraft] is coming from?

R: It wasn't long before I realised you could pre-empt a whole lot of things, just simply by mental imagery / each button [on the console] the realisation was eventually, that each button reflected a particular block of air-space there. As soon as that light flashed, you mentally thought, *'North west'/* or *'block south-west'*. And not only did you identify the block of airspace, but by looking at the strips you were holding there, you could predetermine, to an extent, what was going to come. So you're already turning things inside out. If the call was from [from a particular airspace sector] as you're about to answer it, you're looking at your own outbound track to see what you've got/ You're making and predetermining/ At the same time, [another airspace sector] rings, but, you're not holding any strips on anything from [that sector], so straight away, you take the call that you've got strip-work for.... When you do eventually take [the other call], they might be asking you for a footy score or *'have you seen Jack or Fred'* or *'What's Bill doing these days?*

Nevertheless, together with the temporal experience of the work, there is a rhythm to the work and new controllers build up their experience of 'routine trouble' as well as gaining experience in non-routine trouble over time.

Affective experience of ATC work

Boud and Miller (1996) claim that 'the affective experience of learners is probably the most powerful determinant of learning of all kinds' (p. 17). It is argued that affective experience is significant for all people engaged in productive activity, not just for those who are engaged in activities traditionally defined as learning. This is because as Donaldson (1992, p. 12 in Boud and Miller 1996, p. 17) suggests, we 'experience emotion only in regard to that which matters'. Psychological arousal also is present as affect because the decisions made involve the self and are about issues that matter.

Affect and performance: Presentation of the confident self

It is contended that there is a direct relationship between one's performance and one's affect. Affect is evidenced in an individual's feelings, emotions and moods (Fiske and Taylor 2013). In temporally demanding and complex work, where decisions made are significant and publicly available to others, affect plays a key role in enhancing or inhibiting the level of performance. As was discussed within the temporal experience of work, controllers discipline their voices to project confidence because a positive affect is necessary to maintain optimum work flow and to minimise effort. Detection of negative affect will lead to expressions of doubt about the controller's decisions which will lead to a greater work-load and a diminishing of confidence and thus performance. In ATC, the affective dimension of experience can be identified in the ways in which people talk about their confidence in what they are doing. As discussed in Chapter 3, confidence is collectively held to be a necessary, though not sufficient, element of good performance. A good controller is capable in their performance and confident in their belief in their own capacity to do the job.

I: Confidence is the crux isn't it?
R: Once you lose your confidence that's the end of you. You've got to be confident.

For Bandura (2001), affective processes are at the heart of theories of self-efficacy and control. 'Perceived self-efficacy refers to beliefs in one's capabilities to organise and execute the courses of action required to manage prospective situations' (Bandura 2001, p. 2). Having confidence, in one's own capability to perform under certain conditions and using this as a tool to ease work flow is used in ATC as part of impression management (Goffman 1960). Furthermore, confident people believe in their own agency and keep trying – a necessary quality in work involving the dimensions of sociality, temporality and complexity. Affect, therefore, plays a key role in the smooth and efficient handling of the ATC system.

In ATC, work affective experience requires an individual to take certain things for granted – to have confidence and trust in others, in the equipment, in one's own decisions. A certain level of vigilance and checking are necessary to avoid complacency and thus a mistake. Too much checking and doubt, however, not only slows the performance down but is also likely to lead to more mistakes. Feeling comfortable, somewhat relaxed and confident in one's own decisions enables the controller to operate with a certain level of automaticity and to potentially make more difficult decisions that enable the work to flow more smoothly.

Goffman (1960) regarded behaviour in everyday life as a performance, with many similarities to theatrical performances. In Goffman's terms, the main objective in impression management is to sustain a particular definition of the situation, that is, to behave in a certain way that makes an implicit statement about what is real and important in this interaction (Guirdham 1990). The impression of confidence needs to be conveyed to gain the confidence of others. Establishing an impression about one's confidence thus becomes part of one's self projected to others in The Room – that is, that the controller is confident and thus capable. Conveying confidence as part of impression management will be discussed further in Chapter 7, where it will be shown that believing that one's role includes the presentation of a confident self, limits inquiry and thus both informal and accredited learning.

'Good' controlling is not just controlling aircraft safely and expeditiously, but with 'finesse'. It comes from solving traffic problems quickly and easily – with elegance. Controllers gain an aesthetic pleasure from deriving an elegant solution. 'Because a smoothly operating traffic flow is a beautiful thing'. For controllers, the satisfaction comes from their performance, from undertaking the work fast (expeditiously) and from making decisions that impact upon the air traffic system. However, the impact of an individual's contribution on the ATC system is brief, as the following controller describes.

R: You come to work, you clock-on in the morning and you do your job. When you sign off you're no longer responsible, you're no longer active. There's no work to take home. There will be incidents that happen that aren't reported where you know you performed less than your best and they will worry you after the event. But in general if you are still worrying about any aspect of the job when you drive into the driveway after driving home from work, then the job is starting to get to you and you need to modify your behaviour to adjust accordingly. Because any event that was happening or that you had a part in has come to a conclusion by now, well and truly. Within 10 minutes of walking out the door. And it is either resolved satisfactorily or you'll read about it in the paper. And it's as simple as that. There is nothing you can do. The event had a time life of 10 minutes or a half-life almost. Because the impact of your effects on a system are very much like rapidly decaying radioactive material. So if you had something to do with it, then 10 minutes later it might still be critical if you're on a procedural sector. If you're doing Approach in Perth, a jet has come from 30 miles and landed. So he either managed to land or he didn't. And that's end of story.

For many controllers this lack of continuity of work is regarded as an asset: They describe the amount of money they earn and how they do not take their work home with them. For others it seems that this lack of meaningful involvement leads to channelling their energies elsewhere. There was a period in history where many controllers had second jobs. The need to disengage from one's work, to separate one's self from the intensity of that work would seem to be a feature of High-3 environments. In discussions conducted with ambulance personnel, for example, the same desires are found. The 'system' marches on. As work increases in intensity, the capability to shut off would seem an important means of reducing the possibility of stress and burnout (Dell'Erba et al. 1994).

Invisibility of knowledge work and the valuing of performance

When the input into one's work is intense and the output is not easily visible (as is the case with many jobs involving knowledge work), means of assessing one's contribution of work quality need to be found by the worker.

It should be remembered that no individual controller is responsible for an aircraft from departure to landing (although this used to be the case for controllers operating General Aviation Towers, where private pilots would come and fly for a few hours, departing and landing at the same airport). Even controllers responsible for ensuring an aircraft lands safely have only had the aircraft as part of their responsibility for the final five to six minutes of flight. Although one would expect that controllers would perceive that their work provides a meaningful contribution to society (ensuring lives are not lost), the invisibility of the work results in some controllers believing that their work is not noticed and that they receive little recognition for their work, that is, unless something goes wrong. The following controller discusses their feelings about work at the beginning and ending of a shift.

R: You plug in, start with nothing, plug out, end with nothing. You know, *'what did I do?'* That's where you could feel guilty about the money that you earn, because you think *'oh, I don't produce anything'.* 'Social contribution?' Hmmm [respondent was reading off 'characteristics of good work' sheet].

I: Well, it [ATC work] certainly does that — saving lives is pretty important.

R: Well, that's the interesting part. Like people say *'oh, saving lives'.* We don't actually save lives, you know? Like some people think that they do *'oh, gees, I'm so good, I've helped these people out, they didn't die'* And I think, it's more protecting them, than saving them, you know? / I mean it doesn't, — we're there to provide a job, but it's not, — we're not ambulance officers. We don't go and revive people. We're there to provide a service. In aviation, when it runs well, it's well. But when thing go wrong, they go terribly wrong

For the controller above and others, the work involves making sure the symbols or dots on a radar screen never touch each other. For some controllers, the satisfaction they enjoy from their work involves their level of performance within the ATC system: at how fast and hard they can undertake their work. The valuing of a high level of performance and its display, however, led historically to a particular culture of ATC that was identified by management and the industry as needing to change. As early as 1992, in a review of air safety incidents occurring within Australia Ratner (1992) concluded that

> ATS [Air Traffic Services] officers take considerable
> and justifiable pride in their work and in the qual-
> ity of their performance. Unfortunately, the pre-
> dominant measures they are using today are (1 how
> few breakdowns of separation, or close calls, they
> have had; (2 how much traffic they can handle; and
> (3 how often they can accommodate requests for
> direct clearances and desired altitudes (p. 38).

Structural changes associated with the implementation of STARS
(Chapter 2) were introduced to regulate the decisions controllers can make
in undertaking their work. At the time this resulted in such changes being
regarded as deskilling by controllers, particularly those directly affected
by the change (such as Approach control).

I: What is it about it [your work] that is satisfying? Or what is it about it
that gets frustrating?

R: The satisfaction is in getting a mess of aeroplanes in a string and putting them in a nice little chain and playing with them. It is a skill
thing basically. It is a joy in having the skill and being able to do something like that I guess and recognise that there is a sort of certain
elitism about it. This is why it is so hard. Every Approach controller is
the same, some more so. That's why it is hard to change the culture.
'We have got to slow things down' and all the rest of it. Because everyone
gets a buzz out of doing it and doing it well. When you start nudging
into someone's good feelings about the job because you have pushed
through a lot of aeroplanes, because [what] you are forcing him to do
is back off, that is why you are getting all the deep resentment. I can
understand it because every controller has been there, it is just that we
are all there at different degrees.

Moving aircraft quickly was valued in all sectors, in some more so than in
others (e.g. Approach) where one of the structural elements of the organisation of the work involves the time aircraft spent in each sectors. The
valuing of a high level of performance is shared throughout Centres and
sectors and thus collectively held is part of the organisational culture. The
portrayal of a social identity based on the display of performance will be
further discussed in Chapter 7.

In ATC, like other service industries, where work performed is a service, satisfaction comes, not from being able to view a tangible product,
but from the act of performance involved in the work (Orr 1996). Therefore,
a balance needs to be found between being psychologically involved in
the importance of one's work but not to the extent that one worries too

much about the decisions being taken, and conversely, not to become so detached that one sees the dots on the radar screen as something akin to a pinball game. Managing the psychological self, therefore, becomes an important aspect of work performance.

Social experiences of ATC work

As discussed, the complexity of the air traffic system means that no single individual is fully responsible for the entire duration of a flight. Controllers must work together and with others (e.g. pilots) to ensure the flight proceeds safely and expeditiously. Herein lies one of the paradoxical qualities of the nature of the work. On the one hand, one controller is ultimately accountable for their actions. The focus in performance terms (and in learning for that matter) is on the individual. Controllers are expected 'to take charge' of a situation, to 'take command'. This leads many to feel that it is a highly individualistic job. For example, '*You are* responsible, *you will* do it'. On the other hand, ATC is essentially a collaborative activity despite the work being designed to minimise the need for explicit coordination and cooperation between controllers. The controller negotiates, confers, discusses and on the basis of this information, plans and acts. The performance of one controller influences the performance of another, and indeed, the performance of the entire air traffic system is influenced in this way.

Chapter 2 introduced the notion of how the work was spatially organised. Although sitting at consoles with their backs to The Room, controllers have an awareness of other traffic and of controller's operational styles by virtue of their physical presence (by being adjacent to a console) or by being able to view another controller's airspace on the screen. Each controller's actions are interdependent on others in the system and further those actions are responsive to continuously changing and dynamic conditions. Controllers can make work easy or difficult for the next controller by the way they handoff aircraft. Successful collaboration means being able to understand the requirements of the controller operating the adjacent sector such that the work undertaken by one controller flows smoothly into the work of the next controller taking responsibility for the flight.

Controllers attempt to learn the controlling style of others they might be working with or adjacent to, so that they can know who they can trust and who they need to watch; (i.e. those controllers who might handoff an aircraft that might be close to losing separation in the receiving controller's airspace).

Figures 2.2 and 2.4 in Chapter 2 are a historical representation of the proximity of controllers working alongside one another and the ways in which controllers can view the consoles of others close to them. The figures reveal people working and people watching. Those watching were not idle; they were monitoring the situation in a neighbouring airspace. The person standing using the telephone, might have been taking or placing a call from or to another agent involved in the traffic scenario building up on the screen of the person in front of him. Being able to observe the work of others is important in high reliability organisations as it builds in a level of redundancy into the system.

Synchronising work: Interdependence and the body

Just as the body is used as a resource when work is temporally and complexly demanding, the body is used also as a means of working in relationship with others in The Room. Some controllers reported that they enter The Room earlier than necessary to commence their shift to get 'a feel' for how the work activity is going on that particular day. This enables the controller to get 'in sync' with the traffic prior to commencing duties, and to pick up on the environmental conditions mediating the flow at that time. When controllers commence duty they plug in their headset at the console. Some controllers speak of 'plugging in' to the system as if they are 'plugging in' to the sociality or interdependence of the work – that is, that once 'plugged in' they work, not so much as an independent body, but rather as one connected interdependently to others working the system at that point in time. Part of the sociality of this work involves controllers developing what they describe as a 'third ear' that is, they develop their capacity to undertake their work and listen out for what is happening around them. The controller in the following transcript describes the role of the 'third ear' in a smaller Centre.

I: So you've told that aircraft to call and you're waiting to see...

R: You can hear everything happening in The Room at the time. You've got one ear listening to the traffic and one ear listening to adjacent consoles. And the way you use that in an ATC situation is because you're hearing if your transfer to that person has gone through or not. Once you know it's gone through you can ditch your information on that aeroplane, you're not using hot line, you're not using frequency, you're listening to hear Jack go 'gidday Alpha Bravo Charlie, descend to flight level 7000'.

The third ear becomes an important resource for synchronising temporally and complexly demanding work when working interdependently.

Interdependent sentience

The social experience of work involving aspects such as the third ear, synchronising work and public availability of work, in an environment governed by temporality, complexity and affectivity, lead to a quality important in the collective accomplishment of work that is called in this book 'interdependent sentience'. Interdependent sentience involves the quality of having awareness not just of one's own performance, but of the performance of those working in nearby sectors and of the relationship between those aspects and the air traffic system as a whole. Sentience is present because controllers use all of their senses to gain awareness within interdependent work. The social experience of work involves also an implicit knowing of who is going to be requesting what information. For example, when controllers work a sector together (as happens when the work gets busy) this work is regarded as being performed well when each controller can anticipate the other's requirements to the extent that no explicit requests for information are needed. One can imagine similar situations in other settings reliant on close teamwork, such as emergency services, hospital surgery teams and firefighting. It is contended that interdependent sentience is found in many work settings and particularly in work involving time pressure and imperfect information, as Hughes et al. (1992) suggested when they discuss the social organisation of ATC work.

> The orientation of the individual within the social organization of the work is not primarily to the work as a whole, but rather to the tissue of connections and separations as they fan out from the particular position which s/he occupies…. Though often involving extremes of skill, judgement and coordination, and thus intensive 'work', its smooth accomplishment can render the working division of labour silent and virtually invisible (p. 117).

Interdependent sentience is important in High-3 work, as it is one of the elements that gives High-3 work its reliability, or as Hughes et al. (1992) argue, its 'trustability'.

> Taken as a whole, the system is trustable and reliable…. Yet if one looks to see what constitutes this reliability, it cannot be found in any single element of the system. It is certainly not found in the equipment…. Nor is it to be found in the rules and procedures, which are a resource for safe operation but which can never cover every circumstance and condition. Nor is it to be found in the personnel who, though very highly skilled, motivated and dedicated, are as prone as people everywhere to human error. Rather, we believe it is to be found in the cooperative activities of controllers across the

> "totality" of the system, and in particular in the way
> that it enforces the active engagement of controllers,
> chiefs and assistants with the material which they
> are using and with each other. This constitutes a
> continuing check on their own work and a cross-
> check on that of others (Hughes et al. 1992, p. 119).

The sociality and interdependence of the work is a key feature of ATC work in Australia and of work undertaken in other High-3 work environments and is thus a reason why strategies to enhance continuous inquiry is crucial in enhancing reliability in the work system.

Conclusion

This chapter has focused on the ways in which experience is shaped by the way work is organised. It has reviewed the practice of ATC work and its implications for structuring work experience in certain ways. It has discussed how ATC work is experienced temporally, complexly affectively and socially. ATC work occurs within a 'real-time' dynamic work environment and as such cannot be stopped. This means that the work is sometimes very intense and controllers utilise many resources to monitor and undertake the work, including their bodies and organisational artefacts.

The work involves combinations of decision-making that attempt to maximise traffic flow. The temporal and social nature of ATC work experience means that controllers on some sectors need to spend considerable amounts of time to develop expertise in routine and non-routine trouble. The work is experienced affectively; one's affect influences one's performance (and thus the performance of others). A confident controller will make decisions and act in ways that enable a smooth work pattern to be developed. An underconfident controller can create more work for himself because that lack of confidence is picked up socially resulting in others (pilots, controllers) demanding more information. This in turn can lead the controller to doubt his or her own capability which in turn can diminish performance. ATC work is also experienced interdependently – each controller's actions are interdependent on others in the system and are influenced by others' actions.

The next chapter will discuss the ways in which contexts combine to provide opportunities to make the transition between experience and reflection and thus to begin the process of learning.

chapter six

Reflecting on ATC work

The process of reflection involves awareness of an experience and the focusing on or attending to its salient features. Mason (1993) described this as 'noticing' and Kolb (1984) as 'observing'. Labelling is part of the process of noticing or observing and enables later interpretation and sense-making of observations and experiences (the process of conceptualisation is discussed in the next chapter). Labelling assists in identifying the features of the experience so that they may be useful in the future because they assist in triggering recollection or memory. The conceptualisation process begins when the labels we apply in making sense of the experience (or the data available) begin to fit within a matrix of meaning. The noticing of patterns can sometimes occur at a subconscious level initially (Kolb called this 'apprehension' or 'grasping'), later being fully articulated in words. A particular action, or practice may start to work, may 'feel right' and then later the pattern can be recognised, described and generalised to alternative situations (when it has been 'comprehended' as part of the conceptualising process).

Reflection occurs when people, either individually or in groups, ask themselves questions such as 'what is going on here?' and 'why did this happen?' Reflection occurs socially when people communicate their thoughts and inquiries in relation to these questions to one another. In Chapter 3, it was contended that reflection occurs in cultures through the process of narration and it can also be embedded formally in work structures, such as job roles and tasks. Reflection is constrained when such structures block opportunities for reflective activity to occur. Culturally, the outcomes of reflective processes are generated and shared when, for example, people tell stories. The aim in environments requiring continuous inquiry is to make reflection an intentional, active and constant process for both individuals and groups. This chapter will discuss the influence of contexts on enabling and constraining reflection as part of both informal and formal learning activity.

Complexity of experience and reflection

Observing and checking, and seeking patterns is a necessary part of work activity, though as discussed in the last chapter, too much deliberative and conscious reflection at the console can slow work down and create problems because of the temporally demanding nature of the work. However, as discussed, one of the disadvantages of delayed reflection combined with the automaticity required in temporally demanding work environments is that it can make the deliberate and conscious reflection on the skills being utilised opaque even to the controllers directly involved.

Unfortunately, however, reflection and understanding are also inhibited by collectively held beliefs and norms of practice associated with the belief that ATC is based on ability and that the individual either 'has it' or does not (Chapter 3). In the case where an instructor, for example, believes that good performance is based on ability and what the trainee needs is simply sufficient exposure to build up their performance base, little attention is given to making the processes of reflection explicit or to assisting the trainee (or instructor) to develop understanding of the cognitive process involved. The following controller describes a norm of practice that indicates that little time is given to facilitating the process of reflection in on-the-job training.

I: You wouldn't use notes then. What about pre-planning?
R: I don't use notes. We just say *"Right, whose turn for a brew?"* and sit down
     and we just start working and we take it from there.

Reflective strategies could be useful to enhance instructional skills, as the instructor needs to be able to develop the appropriate communication skills to convey his or her own understanding of the emerging traffic scenario and also to be able to allow the trainee to continue to operate, as the following instructor explains.

> The hardest thing at first (when being an instructor) is to be able to concentrate on what is going on. You literally have to teach yourself to actually keep a full picture of what is happening. *'This has to be done now, he didn't do it. I have got to remember that he didn't do it'.* That is the first thing that you have got to teach yourself. That is very hard – being able to keep [a handle on] what is going on.

Strategies to recapture the complexity of the experience, such as using video, would enable instructors to reflect on the timing of their intervention and its consequences. Changes would be required, however, in existing cultures of practice, because within those communities of practice reflection is not regarded as a valuable form of work activity and is currently not built into routines. The implications of norms of practice and their inhibition of processes of learning will be more fully discussed in Chapter 7.

Affective experience and reflection

As discussed in Chapter 5, the demands of work characterised by intensity and immediacy, risk and reliability require controllers to utilise as many resources as possible. In summary, this includes their corporeal and psychological selves. Just as the body is used to notice controllers' management of the temporal dimension of work experience, the psychological self is also used as a means of reflecting on the work activity. Being confident, for example, assists in creating the right conditions for optimal work activity and also signals to other actors in the system, and to the controllers themselves, their capability in the situation. In High-3 work environments, it is clearly important that individuals have confidence in their own capabilities and in the capabilities of others and that they engage fully in undertaking the work, without worrying about doing so, as the next controller explains.

I: What about learning your own limits?

R: I think that is the same once you, overall, if you are not comfortable with it, if you keep thinking *'Is this going to fit? Is this going to fit? Is this going to fit? Is this going to fit? Phew, it has fitted'.* You are not going to last too long. You have to know that *'yes, it is going to fit'* when they take off and not worry about it. You have got to watch it, but not worry about it. When you know that is going to happen.... that you are at that stage, you don't worry so much about your own limitations or something.

As the above controller indicates, being confident in one's decision and that it will work ('yes it is going to fit') provide the controller with self-assuredness and trust (in oneself). This assists in ensuring the controller does not expend too much energy worrying with the consequence of becoming burnt-out (Dell'Erba et al. 1994). As discussed in the last chapter, confidence needs to be conveyed also to gain the confidence of others. A conclusion that can be reached from the analysis is that a certain level of watching and monitoring one's own performance and the performance of others is necessary, but introspection and critical questioning need to be left for after the performance of the shift.

As was discussed in Chapter 3, confidence is sometimes displayed as a means of impression management and its projection can then become part of the controller's persona, part of the social identity projected to others. It will be argued in the next chapter that the display of confidence and belief about one's capabilities can sometimes inhibit reflection and interpretation because by psychologically investing in the 'can do' belief supporting such performance, little time is then spent reflecting and critically questioning one's own performance.

In learning theory terms, dissonance is a key factor in learning (Brookfield 1993; Argyris 2004). However, for dissonance to be stimulus for learning, the learner must have an 'internal locus of control' and attribute mistakes to something that can be fixed with effort and the correct strategy. If the trainee in the above quotation believes that he is not performing well because he does not have the innate ability, then the trainee's confidence will continue to decline and the trainee will be unlikely to be motivated to develop new strategies. Collectively held beliefs in the role of ability in determining success or failure illustrate the influence of culture on the process of both reflection and conceptualisation. This will be discussed further in Chapter 7, where it will be argued that cultural beliefs in ability and performance inhibit reflection and conceptualisation processes.

Many of the interviewees discussed how the dissonance exemplified by the trainee in the above quotation seems to make matters worse because it further undermines performance. Critical reflection can also be blocked if a jolt or dissonance is experienced that threatens a core aspect of one's identity. In ATC, the collectively held beliefs about the importance of competence and confidence lead some controllers to engage in impression management aimed at covering up or avoiding reflections of their weaknesses.

> You can always pick those people [those who engage
> in bravado and bluff] because, two or three days
> after they've had an incident, they're as right as rain.

In the quotation above, the respondent notes the different pattern of behaviour displayed by someone engaging in impression management compared to someone who does not. A controller who does not engage in impression management would be shattered by almost having had an accident and would spend a great deal of time critically questioning their actions and what lead to such a mistake. Controllers engaging in impression management do not display this form of critical reflection. It is contended that for controllers engaging in impression management where the dissonance created between actual and desired performance is so great their very identities and psychological selves are at risk, reflection occurs at a surface level, if at all. This is the reason why the controller in the above quotation appeared to be 'as right as rain'. In High-3 work environments where reliable performance is crucial and where all mistakes need to be used as resources to improve the system, this lack of reflection needs to be challenged and confronted. Fortunately, from the data collected, it would seem that there are not many controllers who engage in impression management and the social experience of work enables others to monitor such reflective avoidance if it is occurring at the console. Making the investigation of mistakes a socially reflective practice (to be discussed below) is another way to monitor and counter reflective avoidance.

Socially reflexive processes at work

Socially reflexive processes are embedded in both structures and cultures. Socially reflexive processes occur when individuals and groups engage in reflective activity in concert with others. In this section, two aspects of socially reflexive activity will be discussed: the spatio-structural organisation of reflection and socially reflexive cultural practices.

Spatio-structural organisation of reflection

The structural organisation of work establishes two important means of socially reflexive practice. The first involves the spatial organisation of the workspace that establishes opportunities for *peripheral reflection*. The second involves changes in the division of labour that differentiated and integrated job roles and work tasks and established opportunities for *proximal reflection*. These two forms of reflection, and the structures that support them, will be discussed below.

Watching others – watching you: Peripheral reflection in ATC work
The physical structure of the workspace such that work is public provides a valuable means of reflection in two ways: It allows controllers to watch others and in so doing, controllers also know that others can watch them. Peripheral reflection is possible because the technology used in ATC work enables those working at the console to notice how someone else is working. Thus, one can watch, notice and label the patterns of work activity from a distance, from the margins, without having to directly inquire, as the following controller explains.

I: What is your sense of how people either share that they have made a mistake, or ask a question of someone who is more experienced? Or say that they were uncomfortable in a certain situation?
R: There are very few people that will do that [ask]. A lot of them will sort of stand back and watch and file it away without discussing it. It is something you can do with radar technique. A guy can watch and see what is happening without actually having to speak with the guy and ask him how to solve the problem.

The visibility of the work to all those in The Room enables individuals to observe the work of others and thus to benefit through observation how someone else might handle a particular problem or the difficulties someone else gets into. In the next quotation, the controller believes their performance is lifted because of their self-monitoring as well as the monitoring of others.

I: How did you build up that knowledge?
R: I guess, because, you make a mistake, not a serious one, but you let yourself down a bit and I guess you're self-assessing all the time and you wonder *'why am I doing that?'* pretty consistently, and you're doing that because you know people are watching you and you want to perform as well as you can.

The visibility of the work enhances this controller's motivation to be constantly self-assessing. The visibility of the work also enables others to monitor the performance of others, should they be engaging too much in impression management. Knowledge of being observed by one's peers provides a motivation to monitor one's own work and to attempt to perform the work well. The public display of work also supports the constructivist view of learning, that what is regarded as 'good' work practice is not determined by some absolute external criteria but by the norms and expectations of a particular group (Putnam and Borko 1997). These are the work practices that will be observed and emulated by others. What is regarded as 'good' performance is determined by the collective values and beliefs (culture) of the group and will be discussed further in the next chapter.

Of course, the knowledge that one's work is publicly available is also an issue of interest to critical organisational theory because of its capacity to be used for monitoring and surveillance. The issue here is not the need to make work activity private but rather to ensure that structures of surveillance (if used) are themselves visible and the ways in which such information is used is transparent and open to question. The question here that needs to be addressed by organisational designers and workplace learning facilitators is: 'What forms of reflection and observations are possible and how this information is used?' Educative work environments need rich sources of information to be available to individuals and to be shared within groups. However, policies and procedures need to be in place to ensure data are used to enhance learning for educative purposes and not for punishment and control. This issue will be further discussed later in the chapter.

Organising reflection into ATC work: Proximal reflection
In the same way that technology enables work to be publicly shared, so too do other structures of work organisation enable reflection to be organised into ATC work practice. As was outlined in Chapter 2, the industrial relations negotiations introduced a change that had two important aspects: The controllers were required to become multiskilled by obtaining a

current rating on three sectors and also those full-performance controllers were all expected to undertake the role of on-the-job instructor. These changes in roles enhanced opportunities for reflection, by requiring that controllers move out of doing the job themselves, to either learning it again as a trainee or by becoming an instructor of others.

This change in role required the controller to stop working automatically and to re-attend consciously to the task at hand. When the controller becomes a trainee again, he needs to 'shift gears' from acting automatically to acting with a deliberate conscious intention towards learning. Systematically moving from one sector to another and becoming a trainee again is also important for the maintenance of certain generic controller skills.

> There was one person that we had, their skills had actually gone backwards when we moved them onto the next sector because they'd been there [on the old sector] for too long. Their thought processes and all [the skills] they had, had just gone into, how can I put it, auto-pilot, and they were just doing it by auto-pilot. But when they were moved onto the new sector, they had to learn again and it was like starting from scratch again because their ability to learn new skills was much lower than it would be if they were still in that learning process.

This excerpt shows what happens to a controller who has remained on a sector for too long. Some of the skills needed for handling complexity, such as flexibility, and for coping with change had deteriorated. In order to perform well, controllers need to be able to learn new requirements, to modify and adapt their practice and to be able to constantly update their skills. The ability to 'learn how to learn' also becomes an indicator of the controller's flexibility and capacity for rapid skill acquisition.

Moving into the role of instructor is also important because it enhances reflection within work activity, something particularly important in High-3 work contexts. The process of instructing sets up opportunities for controllers-as-instructors to observe and reflect on the job of controlling – something that is difficult to do when engaged in temporally and cognitively demanding work. When a controller is instructing another, they sit or stand behind the person working at the console and watch as the trainee controller undertakes the work. This enables controllers working to have others watching. The practice of watching another do the work of controlling enables the controller to view the work activity from a different perspective. Controllers commented that having to become an instructor forced them to re-examine their own knowledge

base. Stepping back from the job at the console had the advantage of giving the controller a different (wider) perspective.

> I think it helps because you are going back to basics
> for yourself. You probably even in fact improve
> your own performance greatly on training because
> you start calculating things again and find a lot of
> things that you would perhaps put in your little
> 'judgement block' in the back of the mind, that you
> use all the time weren't quite right. Maybe when
> you put them in they weren't quite right. So with
> the trainee you are watching all the time and think-
> ing *"What would I do here? What would I do here?"* so
> that you can come in if he is going to ask *"What do I
> do now"*. You say *"Do this"* instead of having to think
> *"Oh God what do you do now?"* – having it all worked
> out in your mind.

Thus, the process of observing a job – particularly one that has the characteristics of complexity – this requirement to take on the role of instructing,
enhances the controller's own opportunities of learning. Some instructors
and some team leaders commented that the instructional role is sometimes used as a strategy to enhance the performance of the instructor.

I: What impact does having a trainee have on you and the way you do the
   job in terms of your currency?
R: In some ways it improves you because you become more aware of what
   is going on around you and you look at other ways of doing things
   and being forced to sit back and watch it in detail - you analyse it a
   lot more. There is an old saying on Approach *'he's been doing the job for
   a year now; it is time for a trainee so he can really look at the whole thing'*.
   Quite often training officers are selected from the point of view to
   improve the training officer not for the trainee's benefit. There is quite
   a bit of that goes on, and there's a lot of truth in it.

In considering the transcript above, a possible first reaction to this would
be to suggest that the main purpose of becoming an instructor should
be to enhance the trainee's performance. However, it is argued from the
perspective of the whole system of work activity, this would be an unduly
limited view. Performance enhancement aspects of the instructional role
of controllers should be acknowledged within structures of work organisation. Embedding the role of instruction within the job descriptions of
all controllers has had, in this organisation, considerable benefits. Data
collected from the interviews conducted in all Centres and all sectors

revealed that undertaking the role of instructor is a structured task, which improves the controller's own performance because it forces the controller to think about, justify and articulate the bases for the decisions he or she would be making if the instructor was doing the work.

Socially reflexive cultural practices

Organisational cultures also enhance reflection in important ways. The culture of a group is evident in the language (and the stories and myths) used by group members and reveals the values and beliefs that link them to the groups they belong to, or to which they desire to belong. Collective beliefs, values and norms shape the interpretations of experience thereby making particular aspects of those experiences available to group members.

Although reflection was described by Kolb (1984) as an individual process, by drawing on sociocultural perspectives, it is contended that the act of reflecting (through noticing and attending, pattern seeking and labelling), involves interaction with others and with the environment. The process of reflection involves observation and dialogue with others in order to know what to become sensitised to and what meaning can be made from an experience. For example, for collective learning to occur, the act of noticing requires an individual to be able to enter the experience of the other. Salient moments that have resonance for others turn into stories – true and imagined – that project the experience into something that others can share.

Collective memory and the telling of 'war stories'

The cultural practice of sharing war stories was introduced in Chapter 3. War stories are a means of collective reflection because they draw attention to certain features in the experience of the work. Through sharing war stories, the controller uses the vicarious experiences of others to reflect on her own practice and to envisage alternative choices of action, as the following controller explains:

I: Do people learn from other's mistakes?
R: Yes, I think so. Like you say, war stories and things like that. Everyone will tell you *'One day I had two here and they did this and that'*. That's a great way of gaining…. If you can't see it yourself, if you can hear someone else's war story and when that happened you get something at that sort of place you think *'I remember hearing that. I'm not going to stuff up like that'*.

War stories act as a form of collective memory and become resources used to guide action. The features of a war story are labelled by the narrator in a particular way that draws attention to values that are important in ATC work performance making the transfer of those values possible through this form of organisational memory. In the next part of this chapter, four war stories will be discussed to analyse what they reveal about what is collectively noticed and remembered in ATC work. The stories are titled 'The breakdown of the labouring body' (introduced in chapter two); 'One too many airborne'; 'Six in the circuit and the jets are coming in'; and 'The trainee who 'decked' his training officer'. The stories are presented in Appendix 2 verbatim to give the reader a sense of the narrative and they will be discussed here in terms of what they reveal about socially reflex- ive cultural processes. The stories found in the Appendix were selected because they were heard in every Centre and thus represent a mecha- nism of collective memory and information transmittal between groups (e.g. 'Six in the circuit and the jets are coming in' and 'The trainee who "decked" his training officer'). They also contain lessons about important attributes of experience discussed in the last chapter that highlights a par- ticular aspect of the experience of ATC work (e.g. 'The breakdown of the labouring body' and 'One too many airborne').

It is contended that the cultural practice of sharing war stories draws attention to certain important aspects of work performance. The first two stories draw attention to what happens if the temporal experience of work is ignored or overlooked in some way. For example, the story of 'The breakdown of the labouring body' reminds controllers of the need to pace their bodies, lest they deplete them and burn out. Note the age of this story – about 60 years – the message in the story is still salient because managing the impact of the temporal demands upon the body is still an important feature of the work. The story 'One too many airborne' reminds controllers of the consequences of temporally demanding work when a controller takes on more than can be handled. This is the same message in 'Six in the circuit and the jets are coming in' and this story will be examined at length below. The story about 'The trainee who decked his training officer' has messages that were heard in all Centres. This par- ticular war story was often referred to by both instructors and trainees when describing the tense relationship that can often develop between them. Underlying this story is the issue of power and its use and abuse in the instructor–trainee relationship. It highlights the belief that the role of the instructor is often to harangue the trainee to determine if he or she has the capacity to do the job under conditions of duress (a point further discussed in the next chapter since the instructor's conceptual schemas, mediated by cultures, influence their instructional behaviour).

The war story 'Six in the circuit and the jets are coming in' (Appendix 2), for example, is a story about how an instructor allowed a trainee to get into a difficult situation. The temporal demands of the workload had become too much for the trainee and in combination with the work's complexity resulted in the trainee losing 'the picture'. When the instructor had to take over from the trainee, he too got caught up in the details of the situation and 'lost the picture' himself and failed to have an accurate awareness of the impending situation. The story also provides an example of the social nature of the work and the relationships between sectors such as the Tower and Approach. The Approach controller alerted the instructor to the impending situation using the hotline, and the instructor, who had not been given an accurate briefing from his teammates, did not pay sufficient attention to the warning. Having realised the urgency of the situation shortly after the instructor began to act and the Approach/Departures controller then took on a role of cajoling the instructor who was already under pressure 'are you going to be ready for this next one? – Departure – do you want it? Do ya?' The person with the knowledge was absent from the workspace and the person responsible for conveying the information (to do the coordination) had 'an overly relaxed attitude'. The story provides a good example of the consequences of a lack of interdependent sentience. Interdependent sentience, as discussed in the last chapter, is the term used in this Thesis to describe the awareness required of controllers operating under temporally and complexly demanding conditions when carrying out work interdependently. Sentience is needed as controllers use all of their senses to gain awareness within interdependent work, as they develop and maintain awareness not just of their own performance but of the performance of others.

The narration of this 'warrie' draws attention to key features of work activity on a number of levels. First, it draws attention to how an instructor can be 'caught out' and the consequences of an instructor extending beyond his limits, as well as for the trainee 'he saw me as being an Ace at the job and/it sort of came back to him'. The interdependent nature of the work is also highlighted within the team as other controllers were reminded of their responsibilities to one another. 'He never did the crosswords any more, / he realised that that job was a lot more important than he thought it was'. When told in other Centres, the story marks the importance of within-team performance and of the dangers of assuming that a light traffic load does not require the same level of vigilance as any other day of the week 'from then on, people were careful on Sunday mornings just like every other time'.

It is contended that socially reflexive cultural practices such as the narrating of war stories are a form of vicarious learning (Bandura 2001), because the stories shared contain valued elements of work practice that are collectively remembered and transferred. War stories form part of collective noticing and remembering about what can happen. They play a particularly important role in High-3 work environments where typical strategies of experimentation such as trial and error are not available (Weick 1987, 2012). War stories enable other controller's experiences to be shared and form a means of collective reflection because they remind organisational members to notice certain aspects of the work and the consequences when particular actions are not attended to. The marking, labelling and recalling of collectively held values of work performance, and what can go wrong, are then transferred across groups operating in the ATC system. Through war stories, other controllers vicariously experience non-routine trouble and reflect on what needs to occur to avoid that situation. Sometimes war stories become the basis of simulator training sessions, as was the case in the 'One too many airborne' story – so the trainee can learn the importance of recognising when one's limitations have been reached and saying 'no'.

Learning through war stories

As demonstrated in the above examples, people collectively reflect on many things when they engage in the telling of war stories because these stories form part of organisational memory. One's level of performance, developed through a variety of experiences, enables those with the most experience to tell the best war stories (and to participate in this kind of knowledge generation and transmission).

Stories are important because they register, summarise and allow reconstruction of scenarios that are too complex for logical and linear summaries to preserve (such as a set of regulations) (Weick 1987). According to Weick (1987), a system that values stories, storytellers and storytelling is more reliable that a system than derogates these forms of learning, 'because people know more about their system, know more of the potential errors that might occur, and they are more confident that they can handle those errors that do occur because they know that other people have already handled similar errors' (p. 113). Telling stories, therefore, assists individuals to make sense of the non-routine. Through stories, group members accommodate the unfamiliar to their existing experience.

What happens, however, when the wrong messages are conveyed through cultural forms of collective memory such as the narration of war stories? War stories are a valuable means of enhancing reflection if they increase, to use Westrum's (1993) term, 'requisite variety'. The issue here is the need to develop strategies to enhance diversity of stories and to encourage critical reflection to analyse the messages inherent within them so that the values contained can be evaluated to determine whether or not such values are appropriate to continuous learning practice. Just as stories can be used as an aid to learning, however, they can also inhibit learning, since they can reinforce dominant stereotypes and norms of behaviour and thus non-learning. Encouraging critical reflection on the values underpinning war stories would be one means of encouraging changes in practice when those cultures support norms that would inhibit learning behaviours.

Conclusion

This chapter has outlined the ways in which the process of reflection is enabled and constrained within structures and cultures governing work activity. From the analysis and discussion, we can conclude the ways in which structural features such as the physical organisation of the work and its influence on experience (such as its intensity, immediacy and complexity) inhibit reflection and that this occurs in both accredited and informal learning. Other structures, such as those associated with policies of multitasking, enhance reflection because the job roles are predicated on controllers returning to the process of consciously attending to the elements of controlling work. These work structures enable the controller to step outside the work of controlling and to observe another doing the work. In this case, being involved in the task of accredited learning leads to informal learning for the instructor. This particular finding also illustrates the linkage between aspects of accredited and informal learning. An important conclusion for future work organisation is that job roles should be evaluated in terms of what they enable or constrain reflection on experience within the job. Similarly, the design of work environments, for example, the degree to which work is publicly available to others, and the physical spaces available for people to meet and talk are important in enhancing opportunities for reflection within a work setting.

In environments characterised by High-3 work, the telling of war stories provides a cultural means of informal learning as people share experiences, thus enabling others to reflect on those experiences individually and collectively. Telling stories assists individuals to make sense of the non-routine. Through stories, group members accommodate the unfamiliar within their existing experience. Thus, stories provide a means of collective remembering and they enable experiences of an organisational system to be shared vicariously. Interpreting these reflections by identifying patterns and generating others for use in the future and for reframing or re-evaluating experiences is the process of conceptualisation and will be addressed in the next chapter.

chapter seven

Conceptualising ATC work

The process of conceptualisation involves making sense of reflections through the process of interpretation. The new understandings that are generated through reflection can then be applied or generalised to other settings.

In terms of cognitive learning theory, schemas provide a means of organising experiences and are influenced by socialisation (Augoustinos et al. 2014). A 'schema' is a mental conceptualisation or structure which contains general expectations and knowledge about the social world (Augoustinos et al. 2014). Schemas provide a context of making sense of what is noticed and observed. This is because a schema guides identification of what is noticed by providing a context for its meaning. For theorists of social cognition, 'information processing is, ... conceptualised as theory-driven rather than data-driven; that is, it relies on people's prior expectations, preconceptions and knowledge about the social world in order to make sense of new situations and encounters' (Augoustinos et al. 2014, p. 43).

Augoustinos et al. (2014, pp. 36–42) identify four different types of schemas individuals hold relating to: persons, self, roles and events. They are:

- Person schemas which are used to make inferences from the experiences of interactions with other people based on conceptualisations about personality traits
- Self-schemas, which involve the conceptual structures people have of themselves and which guide their own actions
- Role schemas which comprise the knowledge structures people have of the norms and expected behaviours of specific roles in society
- Event schemas, which are used to anticipate the future, set goals and make plans

Later in this chapter, these schemas will be used to demonstrate how two different workplace cultures have arisen based on differing self, person and role schemas. It will be contended that these collective schemas have led to differing conceptualisations of job roles and learning which have in turn led to different norms of work practice. The consequences for both accredited and informal learning in the workplace will then be discussed.

Schemas and conceptualisations about the social world

People use schemas when they make sense of the social world. Schemas work by mediating what people notice, what they look for and how they account for their and others' actions. In the following quotation, a trainee describes her conceptualisation of what happens to trainees when an instructor's schema holds that the trainee does not have the ability to do the job.

I: Can you give me an example of how somebody's life would be different from having done [the training]?
R: Some of them were off work for two or three months, just unable to cope.
I: With training?
R: Yes. And being pushed.
I: Pushed over the edge?
R: Being pushed over the edge, yes. These are people that are, for all intents and purposes, competent. It seems previously, this is only hearsay, that basically turned up the first week they decided whether you are going to make it or not. If you were going to make it they would probably have made it a bit easier for you. If they decided you weren't going to make it, then they set out to prove that you weren't going to make it.

The above represents a person schema. According to the trainee, an instructor perceives that a trainee does not have the traits necessary to do the job (ability) and thus sets out to prove that the trainee is not capable. Good controllers do have special abilities such as being good at mental calculations. As discussed in Chapter 1, however, the collective valuing of the importance of ability as an innate sense can detract from attending to learning the job and instead emphasises developing strategies for detecting the presence or absence of these seemingly innate qualities. Although all controllers interviewed believed that ATC is based on ability, not all believed that it was solely this ingredient that made good air traffic controllers. However, all who were asked could identify other controllers who did believe this and who, as instructors, made it their task to weed out those without The Right Stuff (as discussed in Chapter 3).

In learning theory terms, collective support for a belief in ability is documented in the literature about the role of internal dispositions in expertise (Ericsson et al. 2006). However, research on expertise is equivocal about the degree of importance that should be given to special attributes involved and points instead to the role of motivation and length of practice (Ericsson et al. 2006). Two views about ability are commonly found in the literature (Owen and Page 2010): that there are those who hold an 'entity' view and others who hold an 'incremental' view. An entity view assumes that ability is an internal stable, uncontrollable trait of the personality, a characteristic that cannot be changed. An incremental view on the other hand, suggests that ability, whilst internal to the individual, is also unstable and, therefore, controllable. That is, ability can be improved through effort, controllers who hold an entity view of ability are likely to believe that it is something innate about that controller's personality, and that one's level of ability cannot be changed. Therefore, when operating as instructors, trainees or controllers are unlikely to look for ways to improve a situation and, as alluded to in the quote above, instead will emphasise, not learning and improvement, but strategies aimed at detecting the presence or absence of those abilities.

When people engage in a process of conceptualisation, they begin to categorise their perceptions and those categorisations are mediated by collectively held beliefs and values. Categorising people on the basis of limited information is known as stereotyping (Augoustinos et al. 2014). Stereotyping can be both negative and positive. Both forms of stereotyping are a form of learning, though they are likely to constrain future learning if limited cues are then drawn upon to interpret situations and thus influence future behaviour. If stereotypes have been established and these stereotypes or schemas are reinforced through previous histories of experience, then individuals from particular groups may get a more difficult time in training from an instructor who holds a particular conceptualisation about the job and learning. For example, if an instructor holds entity view of ability and that particular groups of people, such as those from Flight Service officers did not have 'what it takes' then these trainees would have been given a much tougher time in their training, than would be the case if the trainee had an instructor who held an incremental conceptualisation of ability.

In sociocultural learning theory, individually and/or collectively held beliefs and values mediate what an individual learns from an experience because they filter the interpretation of that experience and this is the basis for constructivism.

Ascribing to particular aspects of these beliefs and values becomes group defining and thus part of one's social identity. Augoustinos et al. (2014) make the important distinction between individual identity and social identity. 'Social identity is that part of the individual's self-concept which derives from their knowledge of their membership of a social group (or groups) together with the value and emotional significance of that membership' (p. 98). Social identity is not just an aspect of individual or personal identity, since even asocial descriptions of self subtly depend on particular forms of social organisation (Augoustinos et al. 2014).

As discussed in Chapter 3, collectively held beliefs about good controlling centre around the display of ability, performance and confidence. Membership in ATC depends to some degree on the demonstration of these attributes. Although being confident is absolutely necessary to be able to undertake High-3 work, confidence becomes problematic when individuals begin exuding confidence because they have invested in the collective beliefs listed in Chapter 3 and they project these features as part of their identity – of being a controller who is always 'in control' – as the following quotation illustrates:

I: Do you think there is also a group of people who feign confidence? Who don't actually have it together but….

R: Oh definitely, and I would probably say that's the overconfident people.

I: So in fact overconfidence is in fact bluster?

R: Yeah. Yeah. And you can tell those people straight away like if something goes wrong, or if they have an incident or something, within a couple of days, they're back on their feet, you know what I'm saying?

I: Right 'didn't affect me'.

R: Yeah. And the reason that comes about is because of the culture, the older culture as well. You have to be seen to be *'in control, I can do anything'.*

I: And so therefore, this is my image that I project?

R: Yeah, that's the image

Creating an impression that one has these attributes, of a controller capable, confident and 'in control', illustrates their emotional significance and their value in demonstrating one's membership in The Room. In the quote above, people feign confidence because of a desire to be part of a group that has these qualities and to be perceived as competent and capable. However, when conveying confidence is used as a resource in impression management, the controller, in fulfilling their role schema, is unlikely to display behaviours associated with informal learning practice, such as asking questions, seeking information or asking for assistance. This is because these behaviours do not fit the image the controller is attempting to portray and, therefore, this conceptualisation about one's work and role has a deleterious effect not just on the individual controller's capacity to engage in informal learning through inquiry, but on the entire system.

Ironically, genuinely confident controllers were found to be more open to discussing problems or admitting their mistakes and do not feel vulnerable when doing so.

> It's a lot easier to admit to a mistake if you're arrogant…. If I'm fairly confident about myself, I'm quite

happy to say *'oh I always fuck things up'*, (knowing
that I don't). Whereas, if I'm very under-confident,
I'm not going to admit to many mistakes. If you're
very confident in yourself as a controller, it's very
easy to admit that you've made a mistake, you say
'Yeah, I fucked that up, I made a mistake'.— Because I
know I really know how to do this, but I know also
that I make mistakes and I can't stop making them
and that's not a problem —. It's not going to destroy
me, knowing that I make mistakes'/ [but] there are
so many people out there [now] that are so vulnerable, they don't dare admit they make mistakes. The
thing is, all of us are a bit vulnerable when we say
'I think you might be having a problem here'. In other
words, we're buying into someone else's business.
Now, you notice that we all quite happily do it, generally. Controllers will stick their nose into other
controller's business. The system survives because
people are willing to do that. You have to have that.
[But] Those that are not all that experienced or confident won't do that, so therefore, they become a
burden. They no longer help the system. And not
only that, but if they become unsure, and won't own
up to it readily and won't admit that mistake.

The two quotes above (feigning confidence and being open to admitting
one's mistakes) illustrate some of the paradoxical effects of self and role
schemas involving confidence and its impression management. On the
one hand, the controller who appears overconfident is likely to be perceived as underconfident and to be attempting to use this impression
to cover up uncertainty because exuding confidence is part of a social
identity. On the other hand, a genuinely confident controller will be more
likely to admit mistakes than an underconfident one. The quotes also
reveal differences in opinion about different groups in the room. The
Enroute controller, quoted above about feigning confidence, believes that
it is the 'older' style of controllers (typically found on Approach) that are
perceived to believe they must be in control, and therefore, are unlikely
to admit a mistake; whereas, the Approach controllers being discussed
by the interviewee directly above believe that the Enroute controllers are
unlikely to admit to making a mistake because they feel vulnerable. This
begins to provide some insights into the issue of contested cultures in
The Room and the mediating role conceptualisation (through interpretation, schemas and identity) plays in the process of informal learning in
the workplace.

Groups, collective conceptualisation and contested cultures

As outlined in Chapter 3, there are a number of work groups in ATC and, in part, these groups are established through organisational structures associated with divisions of labour involving the differentiation and integration of job tasks and functions. The way work is organised then, structures work groups in certain ways and thus concentrates work experiences of those work groups in particular ways. Individuals are enculturated into the experiences of the particular work group. In the ATC workplace, different work groups were identified and some of these were summarised in Chapter 3. As highlighted in Chapter 3, controllers define themselves by the kind of work they do (operational/non-operational work), by the type of airspace worked (e.g. Approach, Enroute) and what history of involvement they have had with the occupation or industry before entering The Room. These identifiers assist individuals to develop self-schemas in relation to their peers and to join in membership with others who hold similar beliefs and values.

Working approach and enroute: Contested cultures across 'The Room'

If the metaphor of *ghosts in the machine* holds, then it will be likely that there are going to be different work cultures evident in different work groups based on the way their practice is organised and the way their experience their work, and this in turn will have implications for workplace learning.

Before outlining these differences, it is important to add the following cautionary note. Just as the beliefs and values outlined in Chapter 3 were labelled as exemplars and not necessarily comprehensive of *all* beliefs and values held by controllers about ATC work activity, so too, the following does not purport to represent the conceptualisations of all Approach controllers or all Enroute controllers or indeed all of the conceptualisations about work that a controller who supports the following might have. They are provided here as examples of the ways in which the way work is organised can influence behaviour and thus learning in the workplace for some individuals and some groups.

The experiences of people working Approach and Enroute ATC vary in a number of ways. The most obvious is the difference in the number of years of experience between the two groups. At the time of collecting the main interviews used in this book, most Approach controllers had approximately 15–20 years of experience, having commenced on an Enroute sector and having worked their way around The Room, progressively, moving on to faster, and thus harder sectors until they reached 'the top' position of Approach. Historically, as discussed in Chapter 2, this had also been recognised in pay scales and status. Over that 15–20-year period, many organisational changes have been instituted. These changes included flattening the organisational hierarchy, introducing measures aimed at standardising work practices particularly on Approach. Many Approach controllers also had the opportunity to work in a range of different locations including spending time at outstations around the country. Conversely, controllers working in the Enroute sectors were likely to have had around one to seven years of experience, were more likely to have either always worked in a team or to have spent much of their ATC working life in one, and because of closure of outstations and the centralisation of services, had not had the diversity of experiences of Approach controllers.

The following quotation illustrates the historic ambition of individuals to work their way around The Room. Being able to do the 'hardest' job (Approach) was rewarded by pay scale before the flattening of the organisational hierarchy and the introduction of teams.

I: Tell me about the ego problem.
R: You have got to believe you are the best otherwise you can't do it. One of the differences about Approach is that, until you get there, you are always aiming at a harder job. You have always got something to aim for and you have got something to prove. Once you get there, suddenly you have done it and where do you go? Everything falls flat. Then they start picking on each other.

The references to the desire to do a harder job and having something to prove, reflects the psychological investment these controllers have made in particular beliefs and values about the job illustrated earlier in this chapter and in Chapter 3. Social identity within this kind of work group then is based on one's performance within the group and its comparison with others and this influences the norms of Approach work groups, as the following Approach controller describes.

> We've had girls [Enroute controllers] come over and say 'You guys all hate each other! You yell and scream'. We say 'it's part of the game! We've been doing it for years. It's a big game. It's one upmanship' and / I guess it's just a release mechanism. Everyone does it. It's just the way it is.

This Approach controller is discussing the reactions of Enroute controllers to the norms of behaviour on this Centre's Approach cell. In contrast, a team-based culture, demonstrating an openness to ask for assistance, to seek other's perspectives on work-based problems, is more evident in this controller's work practice than in the norms of practice found in Enroute teams in general. It is contended that this, in part, is a compensatory mechanism given the inexperience of the controller concerned and the uneven level of experience found across the two work groups.

> Doing it the old way, not in the way the job's [now] done, but doing it the old way as in the mentality. Now I believe the culture in the past has been *'I can control everything. I am a super-controller'*. You know? *'I don't care, keep sending me aircraft' 'Oh what about that, you didn't do that?' 'Oh stuff that, they don't want to know about that'*. It's sort of, people have fallen into this repetitive type of self-reinforcement that they think they're invincible. And that's been proven time and time again, people aren't. We all make errors, so the new culture, that I believe is a better culture, that I start seeing but you know, it gets overridden, by these one or two people, is one where *'Okay, it's ATC, [and] it's serious business'* [compared with] in the terms of this older style controller, *'Planes will hit, people will die'*, — very morbid type of attitude. The new culture is *'Okay guys, we've got a job to do here, let's get it done. I'll help you out, no problems, we'll get it done'*. If there's two people on, why should one person work extremely hard, whereas two people could share the workload and get it done?

This quotation suggests a different type of interpretation of the work environment 'Okay it's serious business' – but let us work together to address the problem. This perspective is in contrast to a valuing of individualistic performance. The Enroute controller above describes the kind of conceptualisation found in communication-oriented teams. In such a climate, team members are more likely to ask questions, share ideas, seek feedback about a particular problem.

In contrast, these norms (and their influence of practices associated with informal learning) can be compared with the norms in Enroute teams, as described in the next quotation.

I: Is the learning from your mistakes shared at all?

R: Yes. I think so. I guess it depends on the people's personalities. Some of the older guys probably wouldn't, — that have been around for a long time. That's not fair to say that all of them wouldn't, but there's some people that wouldn't. I think they'd keep things quiet and not say they've made a mistake because they wouldn't want to admit it. A lot of us are new out there and, I know when I first got rated, I'd quite often solve something and I would say '*Do you want to come over here for a tick? I've just had this happen. This is what I did and what would you have done?*' I'd just get stuff that way because how else do you learn? Luckily the people that I work with are all friends, so if they had something going on you could go and look over their shoulder and say '*What's happening here? What are you doing?*' I think you have to be careful who you do that to. If you know the person well enough, it is a great way of learning things.

As stated, the claim made here is not to suggest that all Approach controllers are individualistic and psychologically invested in certain role schemas which preclude, on the basis of the beliefs and values underpinning those role schemas, activities associated with informal learning. Nor is it contended that all Enroute controllers live and work in open communication-oriented teams. The important point to note from this discussion, however, is that beliefs and values, collectively held, permeate group norms and group cultures. The data discussed here suggest that beliefs and values, as part of the conceptualising process, filter what individuals and groups look for in events and shape interpretation and action in response to other events. In so doing, these role schemas become part of the attributes that individuals seek to emulate in order to gain group membership. They are then in turn reproduced by those individuals in ongoing cultures. These beliefs, values and norms are used to account for and justify behaviours and these then influence the degree to which practices associated with informal learning are likely to be part of the norms of work activity. Conceptualisations influence informal learning and they influence the role schemas and practices used by instructors involved in accredited on-the-job training.

Instructional strategies and conceptualisations about learning

According to a constructivist theory of learning, knowledge is a form of interpretation based on a learner's existing conceptions (Augustinos et al. 2014). From this perspective, all interpretation of experience is filtered by what the individual currently understands and believes and so an individual's belief systems are going to influence what that person accepts and learns. This theory holds true not just for learners but also for explaining the strategies instructors use to instruct and what they adopt from instructor training programmes.

In this study, the three commonly held beliefs about what it takes to be a good controller that were identified and discussed in Chapter 3 (ability, performance and confidence). These beliefs also were found to be important in determining how an instructor interprets trainee learning and performance and are linked to three broad pedagogical strategies under-taken by instructors. The three pedagogical strategies have been identified as 'Acting On', 'Working With' and 'Working Against'. Instructors would also use a mix of these strategies depending on their conceptualisations of the nature of ATC work and the trainee.

Beliefs about trainee learning and performance

Given the cognitive complexity of the work, it is not surprising that instructors hold some common conceptualisations about the importance of the trainee's own ability as a key factor in determining the trainee's suc-cess. The importance of ability in ATC learning is reinforced by structural features such as the stringent testing of candidates that occurs during the recruitment phase and is also mythologised into particular cultural arte-facts, such as the Gun controller who can do the work automatically and effortlessly. It is also important to acknowledge that ability must play a significant role in determining success. However, the issue becomes the degree to which the instructor believes ability is the only factor determin-ing trainee success of one factor of many.

If an instructor attributes a trainee's successful performance solely to the notion of ability (and through the demonstration of performance), then there is little the instructor can do except provide a reasonably safe environment for the trainee to find their own way. Instructors who con-ceptualised learning in ATC in this way, talked about the training process as osmosis: learning occurred as a seamless by-product of the everyday work activity and the trainee 'picked up' what was to be learned.

> It is quite different I think to other organisations and institutions. Maybe airline pilots are a bit of a similarity where a co-pilot sits alongside the cap-tain and you would hardly call the captain an OJTI training person. He just flies an aeroplane up and down and lets the other guy have a go every now and again and when he has done enough landings and take-offs and what-not, they say *'Oh yes, we'll make you a captain now.* That sort of, by osmosis, you take it all in and that is a bit of the way we do it too. What you do is you sit in during work sessions and you throw them in the deep end. In the early days when it gets absolutely bedlam and danger-ous, pull them out, and you get in and correct them all again and as time goes on you get less of kicking the trainee out and you jumping in to make every-thing run again.

The instructional strategy described above, of 'learning by osmosis' and 'throwing them in the deep end' uses little direct instruction or feedback. For instructors sitting behind the trainee monitoring the trainee's interaction with people and technology, the role is perceived as an almost passive one. That is in order to assist the trainee to learn, the best the instructor can do is to do nothing. This means that the trainee is performing without the assistance of the instructor and has thus learned. In these cases, during the trainee's learning the instructor tries not to act or to intervene unless absolutely necessary, so that the trainee can learn from his or her own direct experience.

> We are not in a classroom drawing things on boards. I'm just sitting there watching. There is not much more I can do apart from talking to you about it or anyone else like that; or stopping you from taking my licence off me. There isn't really much more I can do apart from trying to give you some confidence. Stop sitting there behind you in a high chair writing notes. That's got to give you some sort of confidence. Just to let you go far enough but able to keep you out of getting into too much trouble. There's not much more can be done. It is just sitting there letting them go and trying to see what is going on in their head and if they are starting to do things your way and they are saying the right words that you'd be saying at the same time, or about that time, then you can settle back and say *'Right, this person's ready'.*

To these instructors, learning is a seamless part of the act of production, gained by allowing the trainee to have as much of a free rein as possible so that they build up their levels of experience and of confidence and in so doing demonstrate their levels of ability through their performance. Successful performance in ATC is not possible with confidence alone; ability and learning through experience must also be present. However, without confidence successful performance is not possible. Instructors conceptualising ATC learning and performance in this way, therefore, believed a major part of their role to be building up the trainee's confidence. Strategies involved in building up confidence include providing opportunities for the trainee to learn through their own direct experience and building the trainee up to encourage confidence in their own capabilities.

> I tend to let them have a pretty free rein on the sector — say what they want. I let them get away with quite a

lot, so far as phraseologies go, but when it comes to separation, naturally there's a line drawn. You don't allow a breakdown to occur. But I'll allow most other things to occur and hope that's a lesson for them at least there, I hope that that develops their confidence, so that they haven't had me badgering them, reminding them of dozens of mistakes, I try not to tell them every time they've done something wrong, 'cause that tends to reduce their confidence. Let them have a run, and if no-body else mentions it, perhaps they'll feel a little bit more confident *'I handled that well today'*. *'He wasn't in my ear'*. The dangerous aspect is they may think that what they're doing is right, so you've still got to pick a couple of the main points, pick a few things each day, that you want improved and you just hope that they don't do everything wrong.

In this case, by *not* engaging in roles typically associated with adult learning (Boud et al. 2006; Brookfield et al. 2006) such as asking questions, providing feedback, making notes to support later discussion, the instructor believes that he facilitates the trainee's confidence. The degree to which instructors conceptualised learning as building on the trainee's ability through exposure and the development of confidence leads to the employment of three broad instructional strategies, described outlined below.

Acting on

Instructors employing Acting On strategies typically conceptualise ATC work as based on an entity view of ability; a need to provide exposure to build up their own reservoirs of experience; and a need to imbue the trainee with confidence. Instructors engaging in a pedagogical strategy of *Acting On* would have the trainee perform the work by both directly intervening and telling the student what to do, or by not intervening at all to give the trainee confidence. These strategies were colloquially called within ATC the 'ventriloquist-dummy' and the 'Mother Theresa' approach to instruction. The strategy of performing the work by telling the student what to do is akin to what Rogoff (1990) found in examining the instructional features of social interactions where the major goal of the activity was either on the teaching or on the accomplishment of the task. ATC instructors utilising an *Acting On* strategy, take the accomplishment of the task as the major goal of the activity, and what the trainee learns, they will learn by themselves.

It's like some guy has his hand up your back, he says *'turn left 370' 'Turn left 370'. 'Take up an easterly*

heading'. 'Take up an easterly heading'. And I'm somebody's mouthpiece and, you're fighting it to a certain
extent. But you know it's the only way you're going
to learn. It's the only way most of us learn anyway.
It's the only way I'll ever learn I know that. And he's
telling you what to do, and you don't like it.

Acting On can also involve no intervention at all. The instructional strategy
in this case occurs by allowing the trainee to do the job by themselves and
for the instructor to display, largely through non-verbal communication,
that they believe the trainee can do the work. The premise underlying this
strategy is that the instructor has faith in the trainee's capacity to perform.
In the following excerpt, the trainee previously had been *Working With* an
instructor who employed the direct intervention 'ventriloquist-dummy'
approach and was now working with another instructor, who employed
the 'Mother-Theresa' approach.

An aeroplane asks for something and they turn
round and look at you and you go.... and [you] look
away. All of a sudden you can tell that they have
got a free hand and they are rapt. They get into it
and they are working away and they are doing it
all. They turn around and smile at you and you go
'Good, good. Has anybody seen the newspaper?' Then
you settle back [pretending to read the newspaper]
and you are still watching. They are rapt and they
are working and they are enjoying themselves and
they are having a ball. I ignore them. I don't take
notes. I think it puts them off. *'If he's not writing
notes I must be going OK'.* He will be watching out
of the corner of his eye - *'he's writing something. Shit!
What have I done'?* It does, it distracts them. So I just
sit down there at the same height, usually resting
on one elbow and looking a bit relaxed. When you
tense up they can tell.

In taking a 'Mother Theresa' approach instructors act on their trainees
by attempting to imbue confidence to allow the trainee to learn themselves. In these cases, the instructor conceptualises learning of ATC as a
process involving the trainee building up experience and confidence so
that they can expand on their given abilities. Such instructors will not use
the adult learning and coaching strategies conveyed in the OJTI training
programme because such strategies are believed to negate the instructor's
purpose (i.e. to give confidence). The process of writing notes, for example,

is believed to sap the trainee's confidence, since the instructor would only be noting what the trainee did wrong. *Acting On* strategies also include the degree to which the instructor uses artefacts in the training process, as the following instructor comments.

I: So it is interpreted as a slight?

R: Even to the extent that trainees will know whether you have plugged in [to] the monitor which means you can't come over the top of them [interrupt their radio transmission] or whether you have gone in live which means you can just talk [to the pilot] if you want to. If you go in the monitor, that is a sign that *'he has got confidence in me, I can do the job'.* If you go in live, *'he doesn't think I can do it'.* It is all very much part of the confidence game.

Acting On strategies can involve direct intervention (telling) or no intervention at all (having faith, believing in) and as indicated above, can involve subtleties such as where the instructor stands and how they monitor the trainee's actions. There are advantages and disadvantages to the employment of *Acting On* strategies. Direct intervention and telling the trainee what they should be doing can become a process of scaffolding as described by Engestrom (2004) where the trainee is able to perform work tasks beyond their own capabilities because they are helped by a more knowledgeable other. However, the advantages of scaffolding will only be achieved if the trainee is assisted to take over the work and to be able to do the work themselves. Similarly, believing in the trainee's capacity to perform the job by allowing them to operate the way they wish has the advantage of enabling the trainee to build up their own experiences. However, in attempting to imbue confidence such a strategy sometimes lead instructors to give the trainee an inaccurate impression of how they are performing and may lead the trainee to believe that certain practices are appropriate when they are not.

We have had a few trainees here where it has sort of been trying to be kinder in training and whatever and the person has got to the end of the thing and when they are still not aware of their shortcomings they have been quite marked [inaudible] in there so you have got to sort of balance that when you are trying to cope with confidence – that you are not giving the person a false sense of security or false confidence.

Instructors who employ the Mother Theresa approach as part of an *Acting On* strategy are likely to appropriate from an on-the-job-instruction training programme for instructors those strategies that support their conceptualisations of how trainees learn best and what the instructor's role is. In these cases, providing encouragement and positive feedback support an instructional purpose to 'believe in' the trainee and so are used to this end. *Acting On* strategies are limited though because the trainee needs accurate feedback about their performance and ideas about how they can improve. Without these elements, the trainee may fail assessment. Encouragement needs to be given to reflection on what seems to work and what is not assisting in progress.

Working with

Instructors who hold an incremental view of ability are likely to believe that their instructional interventions will make a difference to the way the trainee learns and will employ strategies to work with the trainee in facilitating the trainee's learning. In these cases, instructors benefit from involvement in instructional training programmes such as those offered by ASA and they adopt the strategies they learn, as the following controller describes.

> The 'conscious competence' and 'unconscious competence' [idea]. I think that we really needed to know that. I thought that was really the first thing - what a good idea it was and then I realised that it is so much that when you are doing the job for years and you get a trainee, not only do you know what you are doing but you do it automatically.

These instructors will also hold beliefs about the importance of ability, experience and confidence, but perceive that these elements are not the only factors involved in the learning process. These instructors, therefore, use strategies associated with train-the-trainer programmes because they are consistent with their role schema as instructors in facilitating learning in on-the-job training.

I: What about taking notes?

R: I have taken notes during the shift and I have found they have been really handy, especially for the debrief afterwards. Because it is really distracting to a trainee and I don't think beneficial at all to go into a long explanation about how something could have been done better when there is still other things to worry about. It is a very easy trap to fall into I think when you are a training officer because your mind is working at a certain speed, you can see where the situations will resolve, whereas the trainee is still thinking it through. So note-taking is of real benefit. I do a post-shift briefing. The last trainee I had did one every shift and that basically just took the formal '*How do you think you went?*' and then (from what I can remember) the trainee would identify everything I wanted to speak about anyway. We would just discuss how maybe this could have been done.

For these instructors, the adult learning (Eraut 2004) and coaching strat-egies (Hager 2004; Sambrook 2005) conveyed in the on-the-job-training instructor programme discussed in the introduction are employed to increase opportunities for learning to occur. Strategies include observing and assessing the trainee's progress as well as interaction or dialogue with the trainee in an effort to assist the trainee to improve. Strategies include questioning, making notes and providing verbal and written feedback that identifies areas of improvement and gives reinforcement. Instructors adopting *Working With* strategies value being involved in training and derive satisfaction from undertaking an instructional role as the follow-ing controller describes.

> I just find it extremely satisfying - I love it. You get a trainee through and it is a real buzz. I guess in a way, working [in The Room] day-to-day, the only time you ever see.... you never get any.... you can work all day but there is nothing at the end of the day. You work, and the only time you are ever noticed is if you stuff up. There is no *'Oh gee you did well today, that was really good'*, there is only *'Oh you did that wrong'*. Not that you want recognition, but it is not as satisfying as [instructing]. You come in and out every day working and you go home and there is nothing. [in instructing] you have had a part in their career, which is really nice.

Since ASA began investing considerable financial and human resources into on-the-job-instructor training programmes, the pool of instructors adopting *Working With* strategies appears (from the data analysed) to be growing. What the data discussed reveals, however, is that providing instructors with information and tools to use to facilitate adult learning only provides part of an organisational development strategy. What is required also is an understanding of the ways in which existing concep-tualisations of learning and performing – individually and collectively held – enable and constrain acceptance of instructional strategies pro-vided in instructional training programmes.

Working against
Part of the role of any instructor is to make assessments about the progress of a trainee and to evaluate whether or not that progress is satisfactory. On the one hand, an instructor has a responsibility to facilitate or enable trainee learning, and on the other hand, the instructor plays the role of gatekeeper, to ensure that trainees who are not competent do not achieve a licence to operate. The final assessment of job competence is made by

the team leader, though the team leader and instructor do confer during the training period and sometimes this consultation involves other team members who are asked whether they would be happy working along-side the trainee. In the majority of cases, the team leader undertakes the check assessment. In some circumstances, however, the instructor has made an assessment that the trainee is not going to be a competent con-troller. As discussed in the introduction, when controllers hold an entity view of ability and conceptualise good ATC performance as based on this view, then the instructor's role schema is not to facilitate learning, but to develop strategies to detect the presence or absence of ability and to weed out trainees who do not have 'The Right Stuff'. In cases where the instruc-tor has decided the trainee does not have The Right Stuff, the instructor will employ a *Working Against* strategy to show that the trainee cannot do the job and in so doing will set out to 'break' the trainee, as the following controller describes:

I: Have you ever had to talk [a trainee] out of it?

R: Umm.... I've never had to say.... Indirectly you can apply pressure on them, where that pressure then becomes the thing that breaks them./ perhaps if it's getting to the stage that they're not progressing you might have to change tack or you might get to the stage where they have to be put under pressure to make a decision. You then say, *'we're going to combine these positions you've got to be able to run this. It's no easier than it is on a Wednesday morning on one sector, we're going to push you through that right now'*, and you say 'We're going to load you up with work' and you load them up with six hours a day work, the busiest sectors you can, and you come up with the most complex problems and you are bantering [sic] them, you are asking them questions, *'What are you going to do'*, *'What are you going to do'* to the extent they are becoming so taxed that they say *'Is this what it's like all the time?'* Now, no it isn't because you've combined positions that perhaps might not normally be, but you say *'That's no worse than if you were rostered five equivalents of Wednesday mornings in a row in the same week. That's exactly what we're doing. We're making sure that you can do that every single day of the week because we can expect that here, so if it's going to happen on one shift, why wouldn't it happen on another shift? We're making sure you can handle that day in day out'* And they say *'No I can't'* and they might make that deci-sion themselves. The other alternative is that you deliberately distract them and prove to them...

I: To teach them their limitations?

R: And similarly, if you're having a trainee that's having problems, towards the end, you can do the same and you can set them up good and proper. And it's not nice, being a training officer is not a nice job a lot of the time, because people have to face the cold hard facts about themselves.

You do have to, ***in some cases, try and break the person***, and try and prove their weaknesses and say *'well, if I'm not here, is this going to happen again?'* And yeah, if you see the ***slightest inkling*** that that could happen, you have to go for it and prove to them that's a problem.

The role of the instructor as gatekeeper employing a *Working Against* strategy is to locate and weed out any weak trainees who are perceived as not being able to control safely, and this is an important aspect of the job. However, the strategy of *Working Against* uses indirect communication strategies to reveal the trainee's weaknesses to themselves, rather than providing the trainee with direct feedback about their lack of performance. *Working Against* is used when the instructor believes that a key factor of performance is ability and that the trainee does not have the ability to be able to do the job. Whilst it is clearly important to err on the side of caution in assessing the competence of workers in High-3 work environments, the strategy of *Working Against* can serve also to reinforce dominant stereotypes of behaviour (Lois 2001; Kimme 2008; Augustinos et al. 2014). This is a major concern if the stereotyped behaviour supporting a particular culture needs to change. As the quote above indicates, 'the slightest inkling' may be a particular demeanour, or characteristic that does not fit with the dominant values of a particular subgroup culture and may have little to do with actual work practice at all. A controller, who was an ex-Flight Service Officer provides the following account of what it is like to be on the receiving end of someone employing a *Working Against* strategy.

I: Training as testing?
R: Yeah and also *'I'm rated, you're not and I'm going to show you a thing or two my friend'.* And I mucked up in the end. I made a mistake. It was bad for me because it knocked my confidence around. There was no support. This guy was *'Aren't I clever, I tricked you'* sort of thing. And it took me a while to realise that I wouldn't have made that mistake if I had of been able to do things in my own sweet time. I mean it wasn't a dangerous mistake, because he knew that it was going to be okay. But he knew that I had not considered a particular thing and he didn't do anything about it. He pushed me and pushed me *'C'mon, C'mon! Can he have this level? Can he have this level?'* kind of attitude. And in the end I gave it away before fully looking at what I was doing and so, it was very destructive and those sort of people should be.... I mean, I didn't, I should have been the one that was counselled [de-briefed] then, I mean I needed it! But he was the one who definitely needed counselling.

For this trainee, the strategy of *Working Against* caused the trainee to make a mistake that had negative consequences for the trainee's affective experience and caused the trainee to lose confidence. For instructors who perceive their major role to be that of gatekeeper, the adult learning and coaching strategies found in the on-the-job training instructor programme are likely to be rejected because they do not match the instructors own conceptualisations of how the job is learned or support the gatekeeper role.

> A lot of this [on the job training for instructors] course is bullshit. It's all very nice to be caring and sharing, but in our job, it's not caring and sharing. Some people need to be hit around the head with a piece of 4 by 2 because it's the only thing they understand. You can talk to people until you're blue and black in the face. Until you start yelling at them or saying *'for Christ's sake, do something'*, people sit there and allow things to happen around them. You have to take control of it. As an instructor, you have to push, push, push, all the time and that's very difficult with trainees some time. It's a mongrel of a time being a trainee.

The quotation 'It's a mongrel of a time being a trainee' indicates, in part, the difficulties trainees can expect when undertaking accredited learning. This is because their own sense of power and autonomy has been removed (and in the case of *Acting On* strategies that involve an instructor's 'hand up your back') and because they are unable to act competently and confidently. Expecting a difficult time as a trainee can also be accounted for in terms of the widespread use of reliance on solely *Acting On* and *Working Against* strategies. Use of strategies based on *Working With* are increasing, as a result of the resources that have been invested in implementing on-the-job-instructor-training programmes. However, what the data presented here also show is the importance of the ways in which belief systems mediate how much information is adopted from typical instructional training programmes.

This chapter has illustrated how current knowledge and belief systems of controllers serve as critical filters influencing and determining what controllers will accept and learn. Educational research (Gherardi and Nicolini 2002; Hodge 2016) has focused on the role of beliefs and experience in enhancing or limiting opportunities for change. This research suggests that the knowledge and beliefs of instructors must also become *targets of change* if a change in instructional practice is to occur. The evidence in this chapter suggests that programmes aimed at facilitating skills development for instructors need to focus on the existing knowledge and beliefs of instructors and to provide opportunities for challenge and confrontation of those belief systems.

Conclusion

Collective beliefs and values inform conceptualisations that mediate informal and accredited learning. Collectively held belief systems provide evidence of how work activity is conceptualised what is regarded as important or not and what is regarded as learnable or not. Further, given that history is embedded in all activity, collective beliefs and values are often contested. This is because individuals are socialised into different groups, and having had different experiences, build up differing conceptualisations about what is important in the work activity. It has been argued that what is conceptualised about work activity is influenced by structures because these structures differentiate work groups from one another and so concentrate work experiences in particular ways. Possibilities of experimentation and action, as indicated by these findings, will be taken up in the next chapter.

chapter eight

Experimenting and ATC work

Experimentation was described as the activity associated with developing choices and new ways of acting. In terms of Kolb's (1984) model, the process of experimentation provides a link between conceptualisation and experience and, therefore, involves processes associated with thinking (such as envisioning and developing new alternatives) as well as immersion in experience (such as experimenting through doing something in a new way). In this chapter, it will be argued that experimentation can be considered on two axes: intention and the number of actors involved. Experimentation can be intentional (deliberate) or incidental (accidental) and it can occur personally for individuals or can be shared collectively.

Although the rhetoric in organisational literature exhorts leaders to allow organisational members to experiment, and indeed, to celebrate mistakes (Watkins and Marsick 1993), such a strategy is clearly problematic in High-3 work environments because the mistakes can be of such magnitude as to result in loss of life. Typical forms of experimentation such as trial and error are not viable options in in operational (console) work contexts. Moreover, it can be argued that in dynamic and uncertain environments, sole reliance on a method of experimentation such as trial and error would not be viable in any organisation that wishes to remain successful. Therefore, other strategies for increasing the ways in which experimentation occurs and is shared need to be identified.

Incidental and personal experimentation: Learning vicariously

Although the High-3 nature of ATC work leads to structures involving a high degree of formalisation, no amount of formalisation will account for every work situation. In these cases, one common means by which new alternatives are generated is through personal incidental experimentation, or trial and error. Trial and error occurs as individuals engage in work practice and, in so doing, notice that certain strategies seem to work better than others, as the following controller explains:

I: You said before that you then had a learning curve after you got rated.
R: Yes. I think most people do, because you get "let go" and then of course you see situations. When you first get rated you don't do things always

as well as you can but because you do it and you stuff up and it's your
problem. You think *'God I could have done that so much better. I won't do
that again'* and you learn from trial and error I guess. You learn really
quickly because it's your mistakes.

I: You are reflecting quite actively.

R: Yes. Every time you do something you are thinking 'That didn't really
work very well. I could have done it better. Next time I'll do it this
way.' You are constantly doing that all the time.

This quote reveals the close association between experimentation and
the other processes involved in Kolb's learning cycle. The newly rated
journeyman controller ('let go' from the confines of the formal on-the-
job training setting) experiences making mistakes and critically reflects
on how and why these mistakes occurred to generate new alternatives
for future action. In the interviews conducted for this study, newly rated
controllers often mentioned the high degree of incidental (and personal)
reliance on trial and error after receiving their initial rating because they
were responsible for the airspace and, therefore, began to learn from their
own mistakes.

I: Someone has said to me that you don't develop technique until you have
got your rating because they are not your techniques – you are just
using somebody else's.

R: You do. When you are finally rated and you waddle off with a rating
in your pocket [that] is when you really start the learning session. All
that has happened up until that point is somebody has sat with you
and beaten you across your head enough times to keep you out of
trouble. Someone to say *'Yes, he is safe to leave on their own, they shouldn't
bang together'.* And from then on, when you sit on your own, as you do
when you go solo flying, and then you really start to figure out what is
going on. How to do this and that and you make a few prize blunders
and so on.

Incidental and personal experimentation, as the controller above explains,
allows the controller to build up his or her own techniques and 'style' of
controlling – of working out what best works for them in solving the problems presented. When engaging in accredited learning, incidental experimentation through individual observation of practice does not just occur
personally for the trainee but occurs also for controllers and instructors
watching others. As discussed in Chapter 5, the publicly available nature
of the work enables controllers to notice how someone else has solved
a particular problem and this in turn expands the range of alternatives
available to the controller concerned (see, e.g. the section in Chapter 6
on peripheral reflection). Structural policies such as the Enterprise

Bargaining Agreement enable both trainees and instructors to learn other ways of operating by undertaking the associated changes in work role. In the following transcript, the interviewee is describing how the role of instructor expands her range of alternatives by watching the trainee solve a particular problem.

I: Can you tell me about that [the need to be patient with trainees]?
R: Sometimes I think *'I wouldn't have done that'* but it works out very well and I tell the trainee that. I say *'I wouldn't have done it that way, but it worked out really well'.*

For instructors like the one above, new alternatives are generated when they observe a trainee do something not within their own repertoire of experience. This form of incidental vicarious experimentation, through watching another take an alternative course of action, generates increased possibilities for action that can then be adopted by both parties. The degree to which these incidental forms of experimentation may or may not be publicly shared and collectively accessible, however, is an important issue particularly when attempting to develop continuous learning environments and this will be discussed next.

Sharing incidental experimentation

Collectively shared experimentation occurs incidentally when, for example, controllers work in the same team and, in so doing, build up a range of alternatives of experience that can be drawn on by group members. This occurs either because team members work alongside one another on the same shift and over time build up a continuity of (present) experience that can be used as a resource in current activity, or because teamwork enables controllers to draw on the (past) experiences of team members as resources in increasing the range of options available. People working together also develop a (sometimes tacit) understanding of the working preferences of other controllers, thereby enabling individuals to anticipate the needs or work practices of another.

Role of teamwork in sharing incidental experimentation

It is contended that just as other artefacts such as tools and notational systems enhance reflection by making work publicly available, a work structure such as teamwork also provides the opportunity for shared 'referential anchoring' (Resnick 1993) and this has consequences for increasing the range of possible actions available. Teamwork enhances opportunities to share awareness and noticing and thus to generate alternatives for future action because as team members work the console together they begin to develop and share a common history. Furthermore, a shared body of experience enables controllers to feel more comfortable about seeking out the experiences of others in relation to particular problems, so that one inquiring into another's experience might learn a better way of handling a presenting problem. The following controller describes the kind of inquiry-oriented norm of practice that has generated in some teams and the outcomes for shared experimentation.

I: Is that learning from your mistakes shared at all. Is there a camaraderie?

R: My team leader I've got now is actually the person who trained me. I'm really comfortable with him. He knows what I can do and I am more than happy to say to him *'Can you come over here for a tick? What's your opinion of this?'* If you don't ask you sit there and think *'I know that wasn't good, but I don't know what else I could have done.'* Sometimes you have got to ask, but I think some people feel a bit intimidated or think *'he's going to put a black cross next to my name if I ask a stupid question'.* Not many of them would do that I don't think. I think you are more likely to get a tick for asking than a black cross for sitting back and saying *"What if it happens again? What am I going to do?"*

As illustrated above, as a structuring device, teamwork provides the anticipation of a shared referent point – when one member of a team has not experienced a particular past but can draw on the past experiences of others. A continuity of experience enables the shared past of one controller to be used in present conversation with another.

The development of a continuity of work experiences does not just enhance informal learning, as described above, but also assists trainees involved in accredited learning. From the perspective of the trainee, teamwork means that he or she meets with the same people and almost without exception, those interviewed believed this enables a quicker learning curve as trainees become used to the work practices of those surrounding them, as the following controller explains:

I: How do you think teams have influenced on-the-job training? Do you think they have made it better, or have they had any impact at all, or none? Or was it worse?

R: I guess from someone coming in who hasn't been in the culture before, it's excellent because you are working with the same people every day. You get to know the people you work with quite quickly. It's not like you come in every day and see more new faces, more new faces. From that side of it it's good because you get to know the people and they get to know you as well. If your training officer wants a break they are happy to sit in with you that's fine. Everyone sort of gets to know each other quite quickly From that side of it, if you get on a good team it's good, if you get on a bad team, you're in trouble.

As the controller above describes, teams have provided a reasonably stable environment for trainees to learn on the job, provided they are part of a 'good' team. The team enables the new controller to become quickly socialised into the organisational culture of that team. However, at issue here is whether the norms of practice within the group support learning – informal or accredited – or as the controller above suggested 'If you get on a bad team, you're in trouble'. While many of the excerpts relating to teams used in this book have been positive, by identifying the ways in which teams enhance the various processes involved in learning, there are many teams in Air Services Australia (ASA) that do not support an open climate of inquiry – necessary for enhancing both informal and accredited learning in the workplace.

Norms of practice enable or constrain experimentation, and thus learning, and illustrate the influence of collective schemas (Chapter 7). From this perspective, working on a 'good team' increases the 'requisite variety' (Weick 1987, 2012) available to team members in the group because people are happy to share what they know. However, structures such as teams will only enhance requisite variety when other conditions are present. Communication patterns evident within a team will be mediated by the controllers' own individual conceptualisations about their roles (i.e. self and role schemas – Chapter 7) and whether or not group norms support this practice. The culture of communication between workteam members, be they acting as trainees, instructors or working as controllers, influences the degree to which opportunities to learn through incidental experimentation occurs and is shared. It also influences strategies used in both formal and informal learning that have as their basis intentional experimentation.

Intentional experimentation

Experimentation does not just occur incidentally or accidentally but also occurs when, individually or collectively, we set out to deliberately increase options, to try something new and to benefit from those experimental strategies.

Intentional and individual experimentation: The role of culture

Conceptualisations of controllers as 'Lone Rangers', as discussed in the previous section, also limit interpersonal inquiry within informal learning and thus the opportunities for the controller to generate alternative strategies through dialogue. In the following transcript, for example, the controller inhibits the possibilities of increasing his own options for handling a problem, through refraining from asking a team member for help because the norm of practice within his team does not support inquiry.

I: Under what conditions would you ask?

R: There have been times in the past when I've been training or I've had a
question or something I couldn't quite understand and I think *'Would
he know? I'll ask.'* But *'No, I'll pick it up'.* And sometimes you do and
sometimes you don't, until later [and] you think, *'oh God! Is that what
they meant?'* And it can be little things. It can be something you really
should have asked about, / [now] I would ask more because I've been
involved with training. If people don't know you don't know, then
you [the instructor] won't tell them, but whether people who haven't
been involved with training would ask? I don't know. Because I
know I didn't. And there was some stuff I really should have asked
about... [But] *'I don't want to look stupid, I should know that'* / or it's been
explained once, it's been explained twice, I still haven't got it. *'Oh well,
I'll pick it up'.* I'd hoped!

For this controller, the process of intentional inquiry is hindered because
the controller does not wish to reveal his lack of knowledge to another. In
this case, generating alternative courses of action was constrained because
of a belief the controller held that 'he should know'. If one has a role or
self-schema (Chapter 7) that as a competent, capable and confident controller, one should not have to ask others for information or assistance, then
this inhibits the controller's use of inquiry and thus of informal learning.
Furthermore, there is also a possibility that the controller's social identity
(Chapter 3) may be at risk by engaging in inquiry, because such actions
may be at odds with the social identity constructed – on being a confident,
capable, controller leading potentially to impression management. These
role and self-schemas also mediate the level of inquiry that may be used to
generate new possibilities for action in the instructor–trainee relationship.
The following controller describes how different she perceives herself to
be to admit to a trainee that she does not know the answer to a question,
and the consequences for trainees if an instructor 'muddles through'.

I: In terms of your own instructional capabilities, what have you noticed
that you have improved?

R: Improvements.... I'm not sure. Over the years, to me if I don't know
something, it is no loss of face to ask someone else. If I don't understand it properly.... / I am not afraid either when I am training to
try. If I don't know something or I can't explain something [to] I have
to ask someone else. I know a lot of other people just try and see it
through so that the trainee misses out.

I: So the instructor is.... are you just saying that there are things that you are
not sure about so you would go and ask someone else as an instructor?

R: Yes.

I: Whereas another instructor wouldn't do that and so therefore....

R: Yes. It is a bit like loss of face if you don't know the answer to this question. I find that when I am a training officer I probably learn a lot any way. That is one of the things that's why I enjoy training because I learn as well.

For this controller, incidental vicarious experimentation is kept private for many people, presumably because the same role and self-schemas of being competent and 'in control' are operating. This instructor feels comfortable enough with their role to not fear 'losing face' by either admitting that there is something that they do not know or in commenting that the trainee did something they had not previously thought about (see earlier quote). In the first instance, the instructor engages in intentional experimentation by seeking to inquire and increase the options (or understanding) for both themselves and the trainee. In the second instance, incidental vicarious experimentation becomes intentional experimentation when observations about the strategy are communicated and noted as a valuable strategy both could try again in the future. In this example, intentional experimentation is mediated by the degree of self-efficacy the controller feels, an indication of the controller's affective experience of work (Chapter 5).

Intentional and individual experimentation: The role of confidence
Confidence is related to a feeling that one can do the job and succeed with effort. Underlying this feeling is the notion of self-efficacy (Bandura 1997). A high sense of self-efficacy leads the individual to feel confident about trying new things and being open to new ideas. In the interviews, it was found that a lack of confidence led to controllers and trainees not being open to new ideas, nor to communication about experimental approaches that might help them if they tried them out. Self-efficacy when manifested in the form of confidence, then, can limit the communicative relationship between the instructor and the trainee in accredited learning, which will in turn limit the trainee's capacity to benefit from the strategies intended by the instructor because the trainee is not open to trying out new ideas, as the following controller explains:

I: Tell me, I guess, from an instructor's perspective, what is it that students have the most difficulty learning? How do you get them through that?
R: I suppose to sum it up maybe, that the trainee is very worried about what is going on. They seem to get into a little shell and you can't communicate with them very well and the people around them can't communicate very well and that is a part of the job. I mentioned that I have found particularly with [name of trainee] for the first half or more of his training, that he was really in this sort of shell and anything you say, you didn't really think you were getting through to him then.

For this instructor, his capacity to reach out to the trainee and to assist in generating alternatives is hindered by the trainee being concerned about what is going on, thus limiting the opportunities for the instructor to bring to the trainee's attention an increased number of options. The reason for this communication problem is likely to reside in what the next controller described as a lack of confidence leading to the instructor not easily being able to communicate with the trainee.

I: Just focusing in on the instructor-trainee relationship for now, could you give me an example of when things worked really well and an example of when things were really difficult with a trainee?

R: An example of it not working very well, is when you can't make your-self, the importance of a particular point being understood. And the trainee in this case has got predetermined ideas – you can see that. Of how something should happen and it's generally that they've picked it up from other controllers, they've overheard it somewhere and they feel that this is the way it should run. In those cases it's quite diffi-cult to get them to change. You've got another problem too, because it's not only that they've got a preconceived idea, because you could give them this other information and they could try it — but they're a little bit frightened — and they've got this idea. They've seen it work for someone else and they're going to keep pushing the barrow, and they're not relaxed enough or confident enough in themselves to try this new idea. So that's why I think it's important to develop some-body's confidence, as long as they're safe, so that they're willing to try different techniques.

The quote also reveals the linkage between formal and informal learning and its relationship with experimentation and confidence. The instruc-tor in the quotation above subscribes, in part, to the strategy of *Acting On* (Chapter 7), the aim of which is to build up the trainee's confidence because the controller believes that doing so will enhance the trainee's affect and this in turn will lead to a greater openness on the part of the trainee to experiment and try different techniques – as alternatives sug-gested by the instructor. The quote also reveals how the trainee's concep-tualisation of the job or 'event' (see event schemas in Chapter 7) of 'how it should run' limits or inhibits the trainee's capacity for experimentation. To build confidence and to enhance trainee intentional experimentation, the instructor builds a 'protective cocoon' (Hughes et al. 1992) around the trainee to allow the trainee to experiment and to build up his confidence and openness to suggestions. This was also discussed by another control-ler as follows:

I: Did you detect that he was not confident?

R: Yes I did and my approach to that was to try and let him go as much as possible — try and let him experiment as long as things were safe. I didn't mind where he took the aircraft around the sky. I wanted to let him build up his own confidence.

In this case, the controller is likely to hold a belief about the importance of exposure and performance (Chapter 7) and to allow the trainee to have as much of a free rein as possible with the intention that through the trainee's experience, their confidence and performance will increase. In this case, it is the instructor's intention to build up opportunities for experimentation; however, these may not be mutually shared. The intentional sharing of experimental strategies is not just limited by culture alone. Structures are also implicated.

Intentional collective experimentation and the organisation of work
Opportunities for experimentation are both enhanced and hindered by the ways in which work organisation emphasises the temporal dimension of experience. If members have not spent much time in the workplace, and thus have not had the opportunity to build up a range of options for handling problems based on their own experience, then this will hinder their capacity to allow others to develop alternatives and to experiment and also will limit the range of alternative strategies they can share. The consequence of this is that the trainee's options for their own experimentation are then further limited. In the following transcript, the controller is discussing the length of time needed in the job (an Approach sector) before that controller would feel comfortable in taking on a trainee.

I: I was speaking with somebody on Approach and they said they felt like they still didn't have the experience after six years.
R: Yes. The things I would say if you went in there to talk to one of these recently rated guys, and said *'How about taking on a trainee?'* They would say *'You have got to be joking'* because it is a very demanding situation in the position and if you are a new boy on the block, whenever you have a problem you have one way of solving it and you might have two ways of solving it. An experienced guy would have two or three ways of solving it and have lots of other options to make it safe. So he can have a trainee. The trainee can run to the edge and make a complete mess of it and he will be able to go in and rescue the situation. The guy with two or three years' experience hasn't got that. He knows he hasn't got that so if he had a trainee, he would have to take over earlier than the more experienced guy.

In the quote above, the lack of depth of job content knowledge, based on limited work experience, in turn limits the kinds of experiences the trainee can have because an inexperienced controller/instructor does not have the same degree of options or resources available for handling the problems the trainee might encounter. The consequences, therefore, are that the trainee's capacity to try out various options is limited in two ways. First, they are limited directly because the inexperienced controller/instructor would need to take over sooner than someone with more experience would and, therefore, the trainee will not have as much opportunity to try out their own problem-solving alternatives. Second, they are indirectly limited because the instructor has fewer options to share with the trainee in terms of strategies for action. This is of concern to newer controllers acting in an instructional capacity because of what they perceive is a 'dilution of experience' found within different sectors in The Room.

> I think, people training after six months.... I mean it's okay because we are pretty fresh out, we still know our theory but you can teach people all the theory in the world and you can tell them how to do the job, but if you haven't got that experience base to hand on to them, they are really missing out I think. Then they haven't got that experience base. So then when their six months is up, they train and they have got no experience / [and] the level of what people can teach gets narrower and narrower. Whereas if you have got someone who has got years of experience they can pass on that knowledge and then they can pass on.... I think the nice-to-know stuff is getting less and less because all we know is the facts and so that is all we can teach. We can't pass on our years of knowledge.

For some instructors, particularly those in the Enroute sectors, the 'dilution' of experience is a concern because it limits the alternatives that can be passed on to the trainee since, as instructors, these controllers do not have the job content knowledge to draw on. Therefore, controllers need opportunities to develop working alternatives in a range of situations before they feel comfortable taking on a trainee because their own experience level and lack of options is likely to limit the experiences they can allow the trainee to have. The concern expressed by this Enroute controller also signal a consequence for the structural division of labour across The Room, where teams of new recruits are concentrated in the Enroute sectors and teams of experienced controllers are, for example, situated on the other side of The Room in Approach.

A structural change such as the introduction of teams then has resulted in uneven access to the histories of experiences available. For example, a team on Approach, which consists of seven members with on average 15 years' experience each, has hypothetically at least 105 years of experience on which to draw. Whereas, an Enroute team with the same number of members is likely to have between 7 and 30 years of collective experience. Given the importance of longevity of experience as a necessary precursor to effective work, teams where there has been less collective experience have tended to develop different cultures that include openness and inquiry as a means of compensating for the lack of individual levels of experience. In so doing, teams with more communicative norms of practice enhance their 'requisite variety' (Weick 1987) and, hence, their options for handling novel situations. This finding has been found also in work groups in other industries. Shaiken (1996) found that skill acquisition was collectively acquired in a newly automated manufacturing plant because the pressure of production left groups of novice workers with few alternatives but to work together to diagnose problems and to plan solutions. Although Shaiken (1996) provides valuable insights into how groups can support each other and enhance the collective development of skill, more systematic strategies of enhancing 'requisite variety' (Westrum 1993) need to be found and these will be discussed in the next chapter.

Conclusion

This chapter has discussed the ways in which cultures and structures enhance and inhibit opportunities for experimentation. Experimentation involves processes such as trying out new options and generating alternative courses of action through thinking, as well as identifying alternatives through reflecting on experience. Experimentation, therefore, can occur incidentally and intentionally, and it can occur individually or collectively. Work environments based on goals of enhancing continuous learning highlight the need to look for ways to increase the amount of experimentation that is both intentional and shared. From an analysis of the data presented, it can be concluded that within ATC, structures such as teams enhance opportunities for experimentation because they enable the diversity of present and past experience to be made available. However, appropriate cultures that enhance communication need to be in place for learning to be enabled in this way. Both accredited and informal learning can be enabled and constrained by the kinds of communication processes embedded within particular workplace cultures. Moving these forms of incidental experimentation into the future requires strategies of intentional experimentation and implicated here are both structures and cultures. Although all forms of experimentation are important, it is argued that changing organisational environments needs strategies aimed at enhancing continuous learning and is the focus of the next chapter.

chapter nine

Conclusion
Strategies for learning and design

The changes occurring within ATC are indicative of the kinds of transformations occurring within many organisations worldwide. This book has explored the impacts of changing contexts on learning in the workplace. In Chapter 1, it was contended that learning in High-3 work environments is really important because it plays a key role in enhancing reliability and safety in such environments.

An 'educative work environment' is one where there is a 'striving to maximize learning in the workplace through the way work, decision-making, technology and related processes are designed, maintained and redesigned. It includes the structuring and evaluating of work relationships based on their individual and mutual learning and knowledge-creation potential' (Kornbluh and Greene 1989, p. 258).

Educative work environments would support processes of continuous learning when structures and cultures enable the four processes in the learning framework used here. Indicators of continuous workplace learning, then, include the ways in which people in the course of their work, have opportunities for learning structured into their workplace experience and have opportunities to reflect on what occurs in their work practice, to make sense of their experiences through interpretation, to generalise those insights to new situations, and to try out new ideas. Accredited and informal learning involves practices of inquiry such as asking questions, seeking information and gaining feedback. As already discussed in this book, these activities can occur at the level of the self (i.e. self-inquiry), in situated activity with others (e.g. asking questions, seeking feedback, sharing ideas and interpretations) and can be embedded within organisational artefacts (e.g. where these are built into ways of organising). The notion of an educative work environment, then, is one where there is constant striving to create opportunities for continuous learning through focusing attention on the cultural and structural context that support such learning processes. Evaluating workplaces to assess the ways in which they maximise learning requires that we explore and analyse how each of these processes are enabled and constrained within the workplace environment.

These findings support the literature on workplace culture and its role in mediating processes involved in thinking and reflecting (Argyris 2004). Although making meaning from experience is an essential element of learning, it also can constrain learning when such collective beliefs become tacit assumptions whose relevance in a new situation are never questioned or tested (Argyris 2004). In this way, the interpretation of experience through certain collective belief structures and value systems can limit reflective and conceptualising processes and lead to what Weick terms 'socially organized forgetting' (Weick 1987). The findings presented in Chapter 7 support this view. The data showed how accredited learning was sometimes used to 'weed out' those not conforming to dominant norms of practice or stereotypes, and because their performance did not meet with instructors' conceptualisations about what was required to be a good air traffic controller (namely, displaying ability and confidence through performance). Conceptualisations based on collectively held beliefs and values were shown also to influence informal learning, because they, in some cases, inhibited practices associated with inquiry.

Structural changes have also led to changes in cultures with negative aspects for informal learning. The findings revealed how, in terms of cultures, storytelling is a fundamentally important component of learning in complex organisations, but that access to this form of learning has diminished due to the narrowing and splintering of functions, and this has resulted in reduced capacities to learn from them. It is argued that the important role storytelling could play in typical workplace learning programmes has been overlooked. Storytelling as a means of informal learning deserves greater attention in workplace learning programmes (Weick 1987).

Implications of findings for the creation of educative workplaces

What are the implications for practitioners interested in enhancing the possibilities of continuous learning and thus the development of educative work environments? Although these findings are based on a phenomenological study of one (High-3) workplace, and thus their generalisability is constrained, it is contended that these findings could plausibly relate to other workplaces and their relevance could thus be tested in other workplaces.

Contextual mediation of workplace experience

In work environments that are temporally and complexly demanding, that involve considerable psychological investment and that occur with a high degree of sociality, a common way of building up a depth of understanding of the job occurs through direct experience.

Chapter 5 illustrated how job content knowledge in ATC was built up through experience over a considerable number of shifts as controllers build up a reservoir of handling both routine and non-routine problems (Suchman 1996). A depth of job content knowledge is clearly important in undertaking work, particularly work that occurs in High-3 environments, and is consistent with research on expertise (Ericsson et al. 2006). In terms of accredited learning, having access to knowledgeable others with sufficient depth of job content knowledge is an important requirement for successful outcomes. Expert workplace instructors have elaborate ways of understanding content knowledge; similarly instructors with a depth of pedagogical knowledge have developed a variety of strategies for assisting and assessing trainee learning. In many professions, learning content knowledge relevant to the job comes with time or experience in the job together with participation in professional development programmes and this is also the case in the ATC workplace.

As discussed in Chapter 2, however, the ways in which opportunities to gain depth of content knowledge are changing. Much of the learning literature assumes a stable, reasonably unchanging environment in which individuals gain their expertise (Ericsson et al. 2006). As the findings of this book reveal, organisational change processes can have a disruptive effect on opportunities to build up content knowledge, especially in workplaces where acquisition of content knowledge occurs primarily through experience in the workplace. Chapters 2 and 7 showed how a number of changes are reducing the capacity for controllers to build up their content knowledge through experience in many important ways. How individuals gain depth of content knowledge in changing environments has particular implications for facilitators of workplace learning. A depth of content knowledge is important because it provides the foundation for knowledge that is used and shared in both accredited and informal learning. Facilitators of workplace learning, then, need to ensure that workplace instructors have opportunities to build up a good depth of job content knowledge and, if not obtainable directly, that they can acquire it by other means.

Such knowledge could be obtained through a range of strategies aimed at supplementing direct experience, which include field trips, development of training programmes aimed at conveying relevant information, talking with others who have greater experience or through simulation. However, the use of all of these strategies would need to be evaluated in terms of the contextual features of the job. For example, although simulation would appear to be an attractive means of substituting for direct experience, and does provide a valuable means of gaining specific skills, the simulator is not a complete substitute for the development of workplace experience, for two reasons. First, in ATC work, the technology available cannot provide a sufficiently authentic representation of the work practice as that practice occurs in concert with everyone else (i.e. the sociality of experience in the way controllers work in with other controllers and pilots in negotiating routine and non-routine trouble). That is, it is useful to develop specific controlling skills, but it cannot simulate the nature of the interdependence of the work activity. Second, the demands for training fully occupy the simulation equipment that is available and this demand increases as the major Centres centralise work activity and take on a greater number of sectors as part of work consolidation. Thus, while particular strategies such as the use of simulation are an important means of gaining specific aspects of experience, they cannot be relied upon in isolation.

The need for workplace facilitators to attend to the means by which individuals build up their job content knowledge is also important because of the increasing emphasis that now needs to be given to boundary issues as a result of increasingly interdependent and complex environments. The findings presented here suggest that there is a greater imperative for the workforce to understand the implications of their job in relation to the wider organisation and system. It is argued that this will be of particular importance in work organisations in the future, especially those involving knowledge work that is embedded within interdependent structures using tightly coupled, complex and/or automated technologies (Zuboff 1988). Developing sufficient depth of content knowledge, as it relates to understanding the internal and external boundaries of the job, thus becomes an important feature for future workplace learning.

As discussed, depth of job content knowledge may not be available because of the lack of experience of team members or because of a lack of opportunities to learn about the interrelationships between various components of a work system. This raises the question of how, under these circumstances, a workplace learning facilitator would aim to increase the requisite variety available to trainees engaging in on-the-job-learning. One means of facilitating the access to job content knowledge for trainees in changing environments is to make the process of facilitating trainee accredited learning the responsibility of the whole team, so that the depth of experience available within the team can be used as a resource. There are a range of ways in which this could be achieved that are not limited to simply sharing the instructional duties around within the team. In appropriate team cultures (one where inquiry-related behaviours are a norm of practice), team meetings could set aside time for trainees to share their experiences to date and to seek input from other members. Such communicative practices could benefit the whole team because they may also provide new insights for team members, thus enhancing collective memory, or drawing attention to divergent practices that may need to be discussed and addressed. Ensuring that teams comprised members with a range of background experiences also would enhance the job content knowledge that would be available in the team.

Just as accredited learning relies on knowledgeable others in terms of job content knowledge, pedagogical knowledge also has been identified as an important precondition for success (Shulman 2005; Corno and Anderman 2015). Pedagogical knowledge is here used to refer to understanding about how learning occurs and how it can be enhanced and assessed. The level of resources available to enhance instructor development will have a direct influence on the collective development of skill. The level of learning enabled by instructing others will only be as good as the level of resources provided to staff to develop their instructional expertise.

As outlined in the introduction, instructors can either adopt or reject aspects of information presented as part of instructor professional development mediated by their conceptualisations of ATC work. These conceptualisations also inform workplace cultures. Much of the literature on workplace learning has overlooked the ways in which individual conceptualisations and cultures mediate workplace learning programmes, including instructor training (Owen and Page 2010; Augustinos et al. 2014), and this area needs greater attention. It is contended that because of the influence of cultures on instructor conceptualisations of learning, the kind of training programme aimed at enhancing instructor skill also needs to be modified from one based on a receptive–accrual form of information transmission to a cognitive–mediational one (Putnam and Borko 1997). This will be further discussed below with the role of conceptualisation. One aspect that will be important in enabling instructional development will be the encouragement of reflective practice for both instructors and their trainees (James 1997).

Contextual mediation of reflection

In terms of enhancing accredited workplace learning, facilitators need to ensure a variety of means of capturing experience so that it can be reflected upon later. The findings presented in Chapter 6 revealed that in High-3 work environments, work experience is often intense and any reflection is delayed, and this also leads to reflection that is often fuzzy and opaque. Utilisation of audio and video recordings provides a valuable means of revisiting workplace experience and these strategies can be used by trainees and for instructor development. Appropriate workspaces where instructors and trainees can sit and talk are also necessary.

Encouraging reflection as part of work activity has been a feature of many professional development programmes (Shulman 2005) though, as has been discussed, some work cultures may resist engagement in reflective practice because such activity is neither part of their history of experience nor part of their collective identity. Within these kinds of workplaces, structuring work practice such that reflection is built into the accredited learning process may be necessary. Instructors and trainees, for example, could be expected to commence and end the working day with 30 minutes feedback time away from the console (although it is acknowledged that this would put added pressure on other team members to cover this time on the console). Enhanced reflection would also be enabled by improved instructor content knowledge as instructors learned the means of facilitating reflection through, for example, appropriate questioning. The ways in which workplace facilitators are encouraged to develop reflective practices in workplace instructors will be discussed in the next section.

Contextual mediation of conceptualisation

The findings presented in Chapter 7 and throughout this book provide strong support for the learning theory of social cognition and constructivism (Resnick 1993; Putnam and Borko 1997). However, most accredited learning programmes are based on a receptive–accrual model of learning in that learning is assumed to take place through receiving information and skills and understandings are accrued based on this transmission.

The notion of constructivism suggests that all experience and learning is filtered by what the individual currently understands and believes. Constructivism has two forms: cognitive constructivism, which focuses on an individual's internal schemas and mental models for making sense of the world; and social constructivism, which emphasises the role of the social context in shaping what is learned (Augoustinos et al. 2014). Therefore, an individual's current belief systems are going to influence what that person accepts and learns. Theories of constructivism challenge the notion of learning through reception and accrual because individual and cultural beliefs will enable and constrain what is observed, noticed and thus received (Resnick 1993). The findings presented in Chapter 7 demonstrated that the current knowledge and belief systems of the controller serve as critical filters influencing and determining what controllers will accept and learn.

A cognitive–mediational approach to facilitating workplace learning would focus on the existing knowledge and beliefs of participants as they learn and provide opportunities to challenge and confront those belief systems when necessary. The role of the workplace facilitator in this model would be one of mediator of meaningful participant learning. Such facilitators would need to be able to know how to create environments that foster learning for understanding and self-regulation, as well as focusing on methods of assessment that reveal student's thinking. An example presented by Putnam and Borko (1997) of a cognitive–mediational programme encouraged teachers involved in professional development to interview someone who is likely to challenge their beliefs. One option for consideration includes having instructors examine their reasoning with other controllers who hold opposing belief systems. Change can also be encouraged through getting instructors to examine their practical arguments. A practical argument describes a person's reasoning about actions by specifying the rationales, empirical support, and situational contexts that serve as premises for the actions (Putnam and Borko 1997). The assumptions behind such inquiry are that when instructors examine their beliefs with a valued other (e.g. a controller who is well regarded), then there is a critical examination of those beliefs that is likely to lead to a change in thinking, or at least a questioning of previous assumptions. Such introspection is less likely if the information is transmitted via an information–accrual model, particularly if conveyed by someone external to the controller's culture.

While beliefs are important in learning, thinking and knowing, a cognitive–mediational approach needs to be seen as a complementary approach, together with receptive–accrual modes of information delivery. In the workplace studied, ATC instructors need to be encouraged to examine critically their knowledge and beliefs about trainees and their learning. Instructors also need support as they learn new instructional approaches and change their existing understandings. Supporting emerging communities of practice as they experiment and try out new ideas in High-3 work environments will be discussed in the next section.

Contextual mediation of experimenting

The findings presented in Chapter 8 showed how current ways of working and trying of new practices are mediated by existing cultures. These findings support theories of social cognition and constructivism, where what is regarded as 'good' practice is determined by collectively held values, beliefs and norms found within the broader work culture. If new strategies are found to challenge existing conceptualisations, as discussed above, then participants need to be given opportunities and encouragement to experiment with those changed practices in intentional ways and be given support until those practices become the norm. One means of doing this is to provide role models who display the kinds of thinking and practice that support goals of continuous learning, enabling participants to engage in private experimentation as they observe and learn vicariously.

Chapter 8 also showed how opportunities for experimentation are mediated by the structuring of work experience as well as cultures and concluded that strategies needed to be found to enhance intentional and shared experimentation, and that this is important in both accredited and informal learning. Intentional forms of experimentation can be shared by creating emergent communities of practice that are supported as individuals and groups begin to learn new ways of acting and to develop new belief systems that accompany changed practice. Emergent communities of practice can be supported by workplace facilitators through follow-up sessions after accredited learning programmes and these may be face to face or utilise other means of group discussion, such as electronic mail. What the experiential theory of Kolb (1984) points to in this context is the importance of creating environments where individuals can experiment in intentional ways and where those ways can be shared. Other strategies associated with shared and intentional experimentation will be discussed under the section discussing the implications for organisational developers later in this chapter.

Summary

The section has discussed the implications of the findings presented in the book for facilitators of workplace learning. It has outlined a range of strategies facilitators could use to enhance the effectiveness of workplace learning. Most of the examples used related to accredited learning settings. The next section will draw on the findings to discuss their implications for organisational development and the focus here will be on informal learning.

Mediation of learning in organisational design

The impacts of organisational design on individual and group behaviour have been studied over a long period (Hall 2002; Kaptelinin and Nardi 2006; Gherardi 2009). What has been largely overlooked, however, is the influence organisational design has on learning within everyday work practice, though this has recently been given attention, albeit implicitly, within human factors research and that associated with computer-supported collaborative work (CSCW) (Engestrom and Middleton 1996). This is described as 'implicitly' because although informal learning is fundamental to the goals of human factors, it is rarely identified as such. The purpose of this section is to consider the ways in which practitioners of organisational development can create environments that enhance learning within everyday work practice. Practitioners may include those interested in human resource development, human factors, CSCW and organisational studies. It should be noted, however, that the features discussed here were found to be salient within one workplace which had a highly bureaucratised organisational structure. This form of work organisation may not have much to inform organisations with very different structures.

As discussed throughout the book, organisations comprise structures involving formalisation, differentiation and integration as well as cultures that are manifested in collectively held beliefs, values and norms of practice, shared language, and histories of experiences shared through stories and other forms of narrative. It has been argued that these aspects of organisation are critical to understanding how learning in the workplace is enabled and constrained. The implications for each of these aspects of organisation will be discussed: that is, formalisation, differentiation, integration and culture.

Structures of formalisation

A summary of the potential effects of structures on learning, of formalisation, based on analysis and extrapolation from the findings, is provided in Table 9.1. This table summarises the impacts of a high or low level of formalisation on opportunities for each of the processes important in learning: experience, reflection, conceptualisation and experimentation.

The degree of formalisation of work procedures refers to one aspect of the division of labour within an organisation. Formalisation in this book is defined as the degree of rigidity applied to ensuring compliance to certain rules and procedures. For example, a high degree of formalisation would lead to experiences with less job variety (in terms of deciding how the job could be undertaken) and as a result, low levels of autonomy; whereas a low level of formalisation would result in experiences of increased autonomy (Table 9.1). Opportunities for reflection and conceptualisation will be influenced by the degree to which these processes are formally embedded within procedures. Engagement in these processes could, of course, occur without such formalisation. However, it is contended that changes through systematising and sharing insights gained from reflection and conceptualisation on work practice will be enhanced if such processes are embedded within work activity and organisational systems. Having sets of procedures and rules (that are strictly adhered to) limits opportunities for independent and incidental experimentation and thus attenuates one means of obtaining innovation. A work organisation that is highly formalised enables the transmission of the change to be quickly adopted by the rest of the workforce.

Less formalised procedures increase independent action and capacity for experimentation, though transfer of insights gained from such action would be slower than if processes of experimentation, like reflection and conceptualisation, are formalised within organisation structures. In work involving continually changing conditions and higher order thinking, routinisation found in highly formalised procedures can only occur to a certain extent and there will always be situations that cannot be anticipated or formalised into ordered ways of working. In these cases, there is

Table 9.1 Influence of structural elements on opportunities for informal learning: Formalisation

Structural element	Experience	Formalisation		
		Reflection	Conceptualisation	Experimentation
High	Tightly coupled, less variety through low autonomy	Can be included in procedures that formalise work activity in terms of opportunities for reflection	Can include enhanced capacities for conceptualisation if built into procedures	⇑[a] Collective capacity for quick uptake ⇓[a] Experimentation
Low	Increased autonomy			⇑[a] Experimentation ⇓[a] Independent action ⇓[a] Transfer across organisation

[a] ⇑ = increase in; ⇓ = decline in.

often a reliance on culture to ensure compliance and standardisation of ways of working.

Formalisation also can be one means by which the organisation builds in feedback loops for learning at group and organisational levels. This occurs when reflective, conceptualising and experimenting processes are built into procedures. Integrating reflection, conceptualisation and experimentation into work protocols has been the focus of much of the human factors research (Reason 1998; Westrum 1997, 1998). Such procedures need to be part of the organisation's integrative mechanisms if learning is to benefit from any feedback gained through such procedures and this will be discussed in the next section.

Structures of differentiation and integration

Structural differentiation is the process of unbundling subsystem activities separating a specific set of activities from others. Structural integration occurs when efforts are made to coordinate the various activities that are created by structural differentiation. This links the differentiated subunits back together through authority, responsibility and accountability relationships. The major features of structures of differentiation and integration and their implications for processes of learning are summarised in Table 9.2. It also outlines the influences these kinds of organisational structures have on opportunities for the four processes of learning.

When work is highly differentiated, it is specialised. If structures of work organisation are highly differentiated and jobs are very specialised, it will provide opportunities for individuals to learn a lot about one specific area of focus. However, highly differentiated and specialised work will involve less variety (Table 9.2). This enables individuals to build up a depth or continuity of experience that can be brought to bear on a particular, specific problem, and historically this was the case in ATC (Chapter 2). In Chapter 2, the reduction in the specialisation of work tasks through policies of multitasking was discussed. If a job is not specialised but contains multiple roles and/or tasks, it will enable individuals to engage in multiple experiences and, hence, to have greater variety in work activity (Table 9.2) and, therefore, greater opportunities for learning through a variety of experiences. However, as was also discussed, a decrease in the level of differentiation can result in increased variety and a decline in the continuity of time employed on a particular task (Table 9.2). If the work involves substantial variety in terms of tasks and roles, there is a danger that the continuity of experience may be lost as a result of being involved in a variety of tasks and what may occur is a superficial level of experience leading to limited learning or a 'diluted experience' in any one activity (Chapter 8).

It has been argued in this book that being involved in a variety of tasks (high differentiation) may facilitate greater understanding of the work

Table 9.2 Influence of structural elements on opportunities for informal learning: Differentiation and integration

Structural element	Experience	Reflection	Conceptualisation	Experimentation
		Differentiation		
High	⇓[a] Variety ⇑[a] Continuity	Determined by the nature of the work activity	Variety of tasks: multiples may facilitate greater understanding of system	Narrowly defined
Low	⇑[a] Variety ⇓[a] Continuity			Greater opportunity
Integration (e.g. teamwork)	⇑[a] Collective continuity across group ⇑[a] Variety of experience available	Sharing of reflections and conceptualisation possible if members comfortable and open to new ideas		⇑[a] Pool of experiences to draw on, leads to more strategies

[a] ⇑ = increase in; ⇓ decrease in.

system because individuals will have been involved in different work activity and will have had contact with different actors involved in the work system. This variety of workplace experience can aid conceptualisation when thinking about how various job roles interrelate. The findings discussed in this book add to this conclusion by showing how level of background experience is a key factor in enabling a continuity of informal learning.

Structures that support high differentiation may lead to specialist activity and structures that support low differentiation may lead to generalist work activity. In a work organisation where there are many people engaged in similar work activity, there will be opportunities for people to converse with one another and thereby to collectively pool experience. It is argued that the kind of work activity (whether specialist or generalist) needs to be considered by organisational developers in terms of the degree to which the nature of the work experience provides opportunities for the development of deep or superficial understandings of work activity as well as for opportunities to move into other processes of learning (i.e. reflection, conceptualisation and experimentation). This kind of evaluation is largely overlooked within the organisational development literature (Jones 1995; Tosi and Pilati 2011) considering the nature of work activity and the degree to which work performed provides opportunities for processes of informal learning.

Structures of integration and their influence on informal learning

Structures of integration are aimed at coordinating differentiated activities (Hall 2002). It is contended that the ways in which informal learning occurs within an organisation will be determined, in part, by the type of 'integrative mechanisms and lateral linkages' (Hall 2002) built into organisational structures. The distribution of knowledge occurs when feedback loops, for example, are integrated into work activity within and between work groups and divisions. Human factors-related programmes such as crew resource management and team building (Baker, Day and Salas 2006) are aimed at enhancing the knowledge that is shared within groups because they increase the level of integration within groups. What is important is the capacity of groups to learn from the experiences of others. The capacity for individuals to move between work groups of different functions will also be important in learning terms because it enables individuals to bring a diversity of perspectives to a particular work activity through having had differing experiences and, therefore, bringing different reflections and understandings to bear on the task at hand and because it generates systems (or environmental) understanding (Anand, Manz and Glick 1998; Baker et al. 2006).

There are a range of factors which assist in integrating activities and all have implications for informal learning. The size of work groups has been identified in other research as influencing opportunities for collectively pooling experience (Sharma and Ghosh 2007). This is because group size is one factor which influences the degree of homogeneity and diversity within the work group thereby influencing opportunities for learning from different experiences and histories and consequent differences in perspectives based on reflection and conceptualisation. These differences are likely to lead to greater diversity in possible action. The sharing of different experiences through reflection and conceptualisation and the generation of possible actions will only be possible, however, if members feel comfortable to communicate these experiences and ideas or can make themselves heard.

An integrative structure such as teamwork, for example, enhances the range of experiences available within the team by increasing the continuity and variety of experiences available within the group. Being involved in teamwork is likely to lead to increased familiarity and thus increased communication (Sharma and Ghosh 2007), though, as the findings in this book attest, the possibility of creating communication-oriented teamwork depends on individual and collectively shared schemas that govern roles and expectations. Where larger groups have the possibility of benefiting from diversity, smaller sized groups, on the other hand, are likely to have greater opportunities for participants to build up familiarity and trust (Sharma and Ghosh 2007). The findings presented in this book showed how these conditions underpin learning because they enhance openness in communication.

Greater attention, however, needs to be given to how information is shared between groups within the organisation, and with groups external to the organisation. Boundary spanning is, therefore, one means by which the organisation connects itself with the environment.

Mediation of work experience by physical work organisation

One of the key roles of this book has been to map out the ways in which the physical organisation of work influences experience by structuring it in certain ways. It has done so by identifying the dimensions of temporality, complexity, affectivity and sociality as being salient within the workplace studied. These findings support the field of practice associated with CSCW (Heath and Luff 1991; Bentley et al. 1992). CSCW investigates the ways in which technologies structure work and communication with others. Such research demonstrates how the social organisation of work occurs in the interactions between workers and their work practices involving technologies-in-use (Heath and Luff 1991). This book extends CSCW investigation in two ways: by investigating the link between the structuring of work activity and the affective dimension of experience (see earlier discussion this chapter) and by extending analysis of the social organisation of work to structures other than technologies, namely, policies and integrative mechanisms such as teamwork.

Organisational developers and designers can investigate the ways in which the physical structuring of work activity bends and shapes the experiences people have and their opportunities for engaging in, and transitioning to, the other processes important for learning. Work that is temporally and complexly demanding will offer different kinds of experiences and learning opportunities compared with work that is not. Work that involves more risk in public activity with others will engage individuals in affectively different ways compared to work that does not. Work that requires high degrees of interconnectedness with others will shape experiences differently than work where interdependence is not a feature. Investigating the structural features of the workplace for the degree to which such arrangements enable or constrain processes of reflection, conceptualisation and experimentation would seem to provide a valuable tool for organisational developers interested in creating educative environments.

Conclusion

This book has focussed on the ways in which *Ghosts in the machine* enable and constrain learning in the workplace through the ways in which the histories of organising work establish particular structures and cultures. These in turn then mediate experience of both individuals and groups. The basic premise of the book is that the lived experiences of people at work are significantly influenced by their contexts (most commonly conceptualised in organisations as structures and cultures) and that these contexts are in turn reproduced or transformed by people. One of the main aims of this book was to understand how these processes interact so that we may create educative work environments in workplace organisations.

The framework used for analysis begins to bridge the gap between the organisation of work environments and the learning that enables work activity to be achieved successfully. This chapter has outlined some enabling strategies to support the work of practitioners endeavouring to create work environments that are educative.

The book encourages workplace facilitators and organisation designers to use these ideas as a basis for other investigations of the role of social history in work and the ways in which structures and cultures influence learning in the workplaces.

Understanding the nature of organisational change and its interaction with learning will become more important in the future as we move more towards tightly coupled, complex and uncertain environments.

Appendix A: Research

The book is based on research conducted over a 20-year period, which commenced with my own informal learning based on the observations outlined in the introduction. This in turn led to a PhD. Subsequently, I continued my interest in aviation human factors through research and consultancy projects and an ongoing commitment to teaching into a graduate aviation human factors program about organisational change.

The approach taken has been a phenomenological or hermeneutic one (Garfinkel 1986), where emphasis is given to understanding the lived and subjective experience of participants through interviews and sustained engagement in the field.

As discussed, I had been initially involved in providing educational professional development. This gave me the opportunity to talk with controllers in training departments and centres. Some initial interviews were set up through serendipity. Having an hour or two to spare, I would loiter around the training department conversing with those present. Invariably, staff visiting the training department (who had not met me before) would ask who I was or be introduced to me. This often provided an opportunity, for example, for the training department supervisor to suggest that the staff member might like to have a chat – 'be a good opportunity for you to talk about that problem child you had' (meaning a difficult trainee) – and so individuals would find themselves ensconced in a room with me and describing their most recent training experience. Other interviews were established through networking. Interviews also were set up as a result of my sitting-in and observing air traffic control at the console. Controllers, nearby, would ask what I was doing. A conversation would start up, punctuated by air traffic control activity. Comments and concerns were elaborated on away from the console. Sometimes, having had a discussion with one staff member, they would suggest someone else: 'There he is now, I'll just ask him'. Roster breaks were arranged.

Interviews

I have conducted multiple (n = 100) qualitative interviews over this time in four air traffic control centres and two training colleges in Australia (see below). These included interviewing trainees and instructors directly involved in on-the-job training in the following work sections: Approach, Arrivals, Enroute and Tower. Interviews were also conducted with people involved in training departments and others with responsibilities for on-the-job training, for example, team leaders, team training specialists, staff involved in quality assurance and human resources and management personnel. Interviews were also conducted with veteran controllers and specialists who were introducing more technological change into the system. Interviews conducted were between 30 minutes and 3 hours in duration.

A series of prepared questions guided the interview, which included

1. I am trying to develop an understanding of how on-the-job training occurs in ATC. Think back to your last trainee (or a current one). Can you describe what happened? What went (is going) well? What was/ has been difficult? Why?
2. Here is a list of instructional behaviours that occur in typical learning contexts. Do they occur here? How? When? Why (not)? (Explore perceived importance).

Instructional behaviours (shown on a card to the interviewee):

- Pre-planning learning opportunities on each shift
- Providing a pre-shift briefing
- Note-taking during the shift
- Providing a post-shift briefing
- Providing regular feedback of progress, including discussing options for change
- Using some kind of workbook/diary while training
- Other

Coding and Analysis

Interviews were then transcribed and analysed using a software program for qualitative data analysis. Interviews were transcribed for analysis with identifying data removed. The data analysis was guided by interpretational qualitative analysis that begins by first gaining an understanding of the entire collected material. A process described by Tesch (1990) as 'decontextualising' then takes place, which involves segmenting the data into meaningful units (i.e. a segment of text that is comprehensible by itself and contains one idea, episode or piece of information). These segments become the beginning of an organising system or 'pool of meanings' to which the data belong. This assembling is termed 're-contextualisation' and results in categories which are further refined to concepts or themes.

The research design included a number of approaches to enhance the veracity of both the data and the subsequent analysis (Patton 1990; Denzin and Lincoln 2008). These included enhancing trustworthiness through the use of multiple and different sources, methods and theories (Patton 1990) and the interviewing of an extensive number of informants to provide supporting evidence (Denzin and Lincoln 2008). Credibility of the findings was strengthened by prolonged engagement in the field and building trust with participants (Denzin and Lincoln 2008). Reporting the data has included using thick and rich descriptions, which adds to verification (Patton 1990). Dependability of results relates to the issue of ensuring data collected is stable and consistent over time. Dependability was enhanced by strategies already discussed. Dependability was also enhanced by collecting data as part of an iterative process. Dependability and confirmability were enhanced through use of audio-taped recordings and verbatim transcripts. Finally, confirmability occurred through member checks of the data (Denzin and Lincoln 2008).

Three distinct phases can be identified in the data analysis. The first was essentially descriptive. I worked inductively with the transcripts to build up a picture of the worlds of participants. This phase yielded the codes and categories used that included experiences described by instructors and trainees.

The second phase occurred when I realised that the influences of culture and structure were not easily observable in my data. There were some obvious examples, though these provided at best tentative conclusions to be confirmed. To tease out and confirm these tentative conclusions I needed to return to the literature and to the field. This was the time when repeat interviews were conducted to check and confirm my tentative findings with participants on their perspectives on norms of work practice and how these practices were accounted for by them. One of the key features of using qualitative data analysis computer programs is that insights gained in the conceptual development process can be documented through memo writing at particular nodes where data is stored (Ryan et al. 2000). Memo writing has been identified as an important aid in theory-building (Strauss and Corbin 1998).

The third phase occurred when the themes emerging were developed both inductively and deductively, through a dialectical process that pulled between: what I had read (in the literature); what I had heard (in the interviews) and then, further, through changing what I understood through dialogue with members of the organisation, involving them in making sense of the data. From this perspective:

> ... the models by which we represent the world are
> in constant movement between the external and
> internal spheres: from externally visible bodily
> movements and postures to internal sensations and
> feelings and vice versa; from external pictures to
> mental images and vice versa; from verbal or math-
> ematical thought to written text and vice versa.
> Thus models are not exclusively individual and
> private; they are also shared cultural patterns of
> thought and action.

(Engestrom 1994, p. 6)

For the most part, I found the people I had contact with eager and willing
to help. Many were concerned about the activity of training and learning
and wanted to contribute. Others had a particular argument to put about
the training system or air traffic control in general and what was right or
wrong with it. Others wanted me to understand why they believed certain things could or could not be done: why, for example, they thought I
was on the wrong track even asking about training in the first place (since
they believed that the skill of controlling had much to do with innate ability and little to do with learning).

I would like to thank all of the people who helped me to tell this story.

Appendix B: Selection of war stories narrated by controllers

Breakdown of the labouring body

A great war story, and true. He had a morning shift followed by a doggo. And he'd done the morning shift in Sydney tower in 1957. And he went home and had a sleep, but was due at a friend's place for dinner that night. He didn't want to go. He wanted to sleep because he was back in the tower at 11 that night. They were committed to going out to dinner and when he got to his friends place, somewhere around Bondi or somewhere, and aircraft were coming over the top all the time. And he was sitting there listening to these aeroplanes going over and he made several references to 'those bloody aircraft' and he'd just had enough of 'those bloody aircraft'. And after dinner, he got extremely tired and he lay down on the sofa to have a rest. He was going to drive straight to work following dinner and he had this breakdown where he couldn't move. Couldn't get up, just couldn't face it any more.

One too many airborne

I have always learnt from that one about one too many airborne. That was a long time ago now. It must be 20–30 years old. There was the two of them doing Brisbane approach and that is exactly what happened. He just got one too many airborne and he didn't have anywhere to put it [the aircraft]. They reckon he just leant over to the other bloke and put his hand on his shoulder and in his broad Scottish accent, he said 'I've got one too many airborne. I am going to have to come over here' sort of thing. The other bloke was 'oooohhhhh'. You learn from that though and you go 'One too many airborne' and as soon as you get really busy launching all these aeroplanes and they [controllers in the tower] go 'next' and you are going 'where am I going to put him to go? Have I got one too many airborne'?

and you say 'Nothing for you'. It gives you the strength to say to the tower 'Forget it'. That takes a long time to come.

Six in the circuit and the jets are coming in

Most of the war stories we relate to are usually excitement in the workplace or something that was humorous or dangerous or something like that. This is my story. I was training someone. It is how you get caught out in this job. It was Sunday morning. There were a couple of [controllers] on breaks and I was training a [trainee]. So we had one position combined* and a [controller] sitting in a position [who] hadn't been doing it that long. [The controller] was rated on that [sector] – but he had a very casual approach and may have been doing the crossword. And I was the Team Leader so I was training the other trainee there. And the other [controller] was downstairs – having a cigarette or something. So he had to go outside the building.

And the [trainee] I was training, I was giving him an idea of what aerodrome control was like. And I had all this airspace released to me. And there were three or four aeroplanes doing circuit training[†]. Right?

Now [Name of Centre] is an international airport, and usually you don't have aircraft doing circuit training there, but you are allowed to *I* and we had three or four doing circuits – all different types of aeroplanes – like high performance twin engine aeroplanes, jet type aeroplanes and little light 150-type aeroplanes.

And then another one asked to do it [circuit training]. And the trainee I was training, he said, 'oh, that's probably too much'? and I said, 'no it's Sunday morning, it's fine'. You know 'bring it on in'. And an aircraft came back from the training area, another little light aeroplane, and he too wanted to do circuits, and I said, 'yeah that's fine'. And [the trainee] *said* 'oh, but that means we'll have six aeroplanes doing circuits'? And as he did that, the hot line went from approach, and the approach controller downstairs who was a bit of a cynical bloke was querying this. So I said, 'It's okay, we can have that many in a circuit, as long as there's no jets'. And the approach controller asked me about another aircraft that was coming in and I gave 'unrestricted' and then hang up.

---

* When the workload is slow, two sectors can be combined to create one sector, managed by one controller. When the work traffic becomes heavy, the sectors arc 'de-combined' creating two sectors of smaller airspace and thus more manageable workload. Sometimes sectors are combined to give the trainee or controller exposure to a higher workload than normal, though in this instance, the sectors were combined due to the light traffic load.

† Aircraft undertaking circuit training are practicing takeoff and landing, without stopping. Typically the pilot will 'land' by touching the wheels of the aircraft on the tarmac and then taking off again. These are generally light aircraft and they maintain their own separation standards by visual means.

And then it got a bit busy. And the trainee, the trainee who was doing it, he'd lost the picture, so I took over. As I'm doing it, this other controller with the very casual approach to things, he walks up [from being outside] the tower and says 'Hey buddy, what are you going to do with that [aircraft] out there with all these other aeroplanes in the circuit'? And I said, 'That's alright, I'll just put [that pilot] in at 1500 feet and drop him through the two Cessna's out there'. And [the controller] says, 'Ah nup – [name of narrator] – That aircraft is a 747'!

And I looked up and there's this big tail fin, coming around the corner. And this other [controller] had put this [aircraft] above my circuit training, in a non-standard flight path, and hadn't told me about it. The controller had written it on the strip. The controller who was doing the crossword [also] hadn't told me, and we all turned around, and here was this big fin – it was just like out of that movie 'Flying High', this big fin coming around the corner.

And for the next 25 minutes – well, just at that time, I'd said to them [approach/departures] 'let me know if there's anyone who's going to push back'. And [I looked and] there's a Cathay 747 taxiing down for a backtrack for the runway.

That aircraft is going to Hong Kong. And then I looked on the screen. Because it had been so busy, I had the screen down small, and I blew the screen out to see if there was anything else coming, and there's [another] two jets coming straight onto approach. So for the next 25 minutes, no one said anything, 'cause I just told 'em all what to do'.

I had six light aeroplanes, that were meant to be doing circuit training – I put them all onto holding points. And all the time this approach/departures controller downstairs says 'are you going to be ready for this next one? – Departure – Do you want it? Do ya'? He kept on interrupting, and I was trying to talk all the time. It was just ah….

We were talking about how sometimes, people take more than they should, and I had done that. I was caught out and I was supposed to be an experienced controller training someone and so for the next 4 months, every time a jet came into [name of Centre] and it was Sunday morning, every controller on approach/departures would hit the line and say 'remember you can only have six in the circuit when you've got jets coming in' and then hang up.

They all did it to me. But from then on, people were careful on Sunday mornings just like every other time because my story got told a number of times.

But that's the thing. It's usually related to something that people had a reasonable excitement or had been caught out. And I had been caught out. And that's the way you learn.

That trainee who was training. He saw me as being an Ace at the job and he saw me get caught out. It sort of came back to him. And the [controller] who didn't coordinate – the [aircraft] who was coining around the corner at me, I actually counselled him over that. 'I may have had too many aeroplanes, but you're meant to tell me what's going on'. From then on, he never did the crosswords any more, (because, he [had thought he] didn't have anything to do because all the circuit training doesn't get coordinated). But he realised that that job was a lot more important than he thought it was. And he got caught out.

The whole thing was, everything was safe, it was just that a whole lot of aeroplanes, that weren't getting told that they were getting delayed, would have got delayed. And I just stood them all out of the way, and got all of the jets in, but these other pilots, who had gone up to do circuits, they were at a fair disadvantage. They were going around in circles over different holding points until I could bring them in.

And no one complained ('cause they could hear me shouting). They are the usual type of warrie – people learn from them.

The trainee who 'decked' his training officer

It is one of the great war stories of [name of Centre] ATC isn't it? Yes. The guy [trainee] that hit [the instructor] is a really nice, gentle, quiet sort of bloke who would do anything for you. The bloke that got hit, I used to work with him in [centre's name]. An arrogant little bugger and nobody liked him anyway. Nobody was surprised when he got hit and everybody thought 'fair enough'. They should never have put the trainee. I don't know, maybe they figured that [the trainee] could handle [the instructor's] arrogance and the way he is. In the end [the trainee] cracked. [The instructor] was just one of those people who would just be nagging and going on and on and on about things that weren't really important. [The trainee] finally cracked. It was all very funny. How [the trainee] kept his job is beyond most people in the ATC.

They were live. They were sitting there doing approach. And you deck someone which means that logically, the controller is out cold, the trainee's in no fit state to do it, and there are aeroplanes going everywhere. Who is going to take over? The guy that is in charge (the boss) had to jump in and take over. Who is going to look after the bloke who just got knocked out? Who is going to sort the bloke out that hit him? All this sort of trouble. It was just so much. If it had been anybody else, they would have been sacked but he was one of the people who was mates with all the people in the right places. Not because of that, it was just because he liked having a beer with people and he was friends with everybody. It saved his bacon.

References

Airservices Australia. 2016. *2015–16 Annual Report*. Canberra, Australia: Airservices Australia.

Alasuutari, P. 1995. *Researching Culture: Qualitative Method and Cultural Studies*. London, UK: Sage.

Alvesson, M. 2012. *Understanding Organizational Culture*. London, UK: Sage.

Anand, V., C. C. Manz and W. H. Glick. 1998. An organizational memory approach to information management. *Academy of Management Review* 23:796–809.

Appleton, K. 1996. Using learning theory to guide reflection during school experience. *Asia-Pacific Journal of Teacher Education* 24:147–58.

Argyris, C. 2004. Double-loop learning and organizational change. *Dynamics of Organizational Change and Learning* 389–402.

Augoustinos, M. and I. Walker. 1995. *Social Cognition: An Integrated Introduction*. London, UK: Sage.

Augoustinos, M., I. Walker and N. Donaghue. 2014. *Social Cognition: An Integrated Introduction*. London, UK: Sage.

Baker, D. P., R. Day and E. Salas. 2006. Teamwork as an essential component of high-reliability organizations. *Health Services Research* 41:1576–98.

Balthazard, P. A., R. A. Cooke and R. E. Potter. 2006. Dysfunctional culture, dysfunctional organization; Capturing the behavioral norms that form organizational culture and drive performance. *Journal of Managerial Psychology* 21:709–732.

Bandura, A. ed. 1997. *Self-Efficacy in Changing Societies*. Cambridge, UK: Cambridge University Press.

Bandura, A. 2001. Social cognitive theory: An agentic perspective. *Annual Review of Psychology* 52:1–526.

Benner, P., A. Tanner and C. Chesla. 1996. *Expertise in Nursing Practice: Caring, Clinical Judgement and Ethics*. New York, NY: Springer Pub Co.

Bentley, R., J. A. Hughes and D. Randall et al. 1992. Ethnographically-informed systems design for air traffic control. *Paper presented at the ACM 1992 Conference on Computer-supported Cooperative Work*, Toronto, Canada.

Billett, S. 2016. Beyond competence: An essay on a process approach to organising and enacting vocational education. *International Journal of Training Research* 14:197–214.

Billet, S. 2002. Workplace pedagogic practices: Co-participation and learning. *British Journal of Educational Studies* 50:457–81.

Blaka, G. and C. Filstad. 2007. How does a newcomer construct identity? A socio-cultural approach to workplace learning. *International Journal of Lifelong Education* 26:59–73.

Boud, D., P. Cressey and P. Docherty. 2006. *Productive Reflection at Work: Learning for Changing Organizations.* Oxfordshire, UK: Routledge.

Boud, D. and H. Middleton. 2003. Learning from others at work: Communities of practice and informal learning. *Journal of Workplace Learning* 15:194–202.

Boud, D. and N. Miller. 1996. *Working with Experience: Animating Learning.* London, UK: Routledge.

Boud, D. and D. Walker. 1993. Barriers to reflection on experience. In *Using Experience for Learning,* eds. D. Boud, R. Cohen, and D. Walker. Buckingham, UK: Open University Press.

Bredo, E. 1994. Reconstructing educational psychology: Situated cognition and Deweyian pragmatism. *Educational Psychologist* 29:23–25.

Brookfield, S. 1993. Self-directed learning political clarity and the critical practice of adult education. *Adult Education Quarterly,* 43(4), 227–242.

Brookfield, S. D., T. Kalliath and M. Laiken. 2006. Exploring the connections between adult and management education. *Journal of Management Education* 30:828–839.

Brown, M. L., M. Kenney and M. J. Zarkin, eds. 2006. *Organizational Learning in the Global Context.* Aldershot, UK: Ashgate Publishing, Ltd.

Breul, C. 2013. Language in aviation: The relevance of linguistics and relevance theory. LSP Journal-Language for special purposes, professional communication, knowledge management and cognition 4.

Bureau of Infrastructure, Transport and Regional Economics (BITRE). 2016a. *Domestic airline activity. Statistical Report.* Canberra, Australia: BITRE.

Bureau of Infrastructure, Transport and Regional Economics (BITRE). 2016b. *International airline activity. Statistical Report.* Canberra, Australia: BITRE.

Cameron, K. S. and R. E. Quinn. 2005. *Diagnosing and Changing Organizational Culture: Based on the Competing Values Framework.* San Francisco, CA: John Wiley and Sons.

Chi, M. T. H., R. Glaser and M. J. Farr. 1988. *The Nature of Expertise.* Hillsdale, NJ: Lawrence Erlbaum Associates.

Child, J. 2015. *Organization: Contemporary Principles and Practice.* Chichester, UK: John Wiley and Sons.

Cole, M., Y. Engestrom and O. Vasquez, eds. 1997. *Mind, Culture and Activity: Seminal Papers from the Laboratory of Comparative Human Cognition.* Cambridge, UK: Cambridge University Press.

Collin, K. 2006. Connecting work and learning: Design engineers' learning at work. *Journal of Workplace Learning* 18:403–413.

Collin, K. 2008. Development engineers' work and learning as shared practice. *International Journal of Lifelong Learning* 27:379–397.

Collinson, D. L. 1992. *Managing the Shopfloor: Subjectivity, Masculinity and Workplace Culture.* Berlin, Germany: Walter de Gruyter.

Corno, L. and E. M. Anderman, eds. 2015. *Handbook of Educational Psychology.* New York, NY: Routledge.

Czarniawska, B. 1997. *Narrating the Organization: Dramas of Institutional Identity.* Chicago, IL: University of Chicago Press.

Dell'Erba, G., P. Venturi, F. Rizzo et al. 1994. Burnout and health status in Italian air traffic controllers. *Aviation, Space, and Environmental Medicine* 65:315–322.

Denzin, N. K. and Y. S. Lincoln, eds. 2008. *Collecting and Interpreting Qualitative Materials*. Los Angeles, CA: Sage.

Dewey, J. 1933. *How we Think*. Washington DC: Regnery.

Dewey, J. 1938. *Experience and Education*. London, UK: Collier Books.

Dixon, N. 1999. *The Organizational Learning Cycle: How We Can Learn Collectively*. London, UK: McGraw Hill.

Donaldson, M. 1992. *Human Minds: An Exploration*. London, UK: Penguin.

Dyrud, M. A. 1997. Learning styles. *Business Communication Quarterly* 60:124–134.

Ellinger, A. D. and M. Cseh. 2007. Contextual factors influencing the facilitation of others' learning through everyday work experiences. *Journal of Workplace Learning* 19:435–452.

Endsley, M. R. 1994. *Measurement of Situation Awareness in Dynamic Systems*. Lubbock, TX: Texas Tech University.

Engeström, Y. and International Labour Office. 1994. *Training for Change: New Approach to Instruction and Learning in Working Life (p. l)*. Geneva, Switzerland: International Labour Office.

Engestrom, Y. and D. Middleton. 1996a. *Cognition and Communication at Work*. New York, NY: Cambridge University Press.

Engestrom, Y. 2001. Expansive learning at work: Toward an activity theoretical reconceptualization. *Journal of Education and Work* 14:133–156.

Engestrom, Y. 2004. The new generation of expertise. In *Workplace Learning in Context*, eds. H. Rainbird, A. Fuller and A. Munro, 145–65. London, UK: Routledge.

Eraut, M. 2004. Informal learning in the workplace. *Studies in Continuing Education* 26:247–273.

Ericsson, K. A., N. Charness, P. J. Feltovich and R. R. Hoffman. 2006. *The Cambridge Handbook of Expertise and Expert Performance*. New York, NY: Cambridge University Press.

Fine, G.A. 1996. Justifying work: Occupational rhetorics as resources in restaurant kitchens. *Administrative Science Quarterly* 41:90–115.

Fiske, S. T. and S. E. Taylor. 2013. *Social Cognition: From Brains to Culture*. London, UK: Sage.

Flin, R. 1998. *Sitting in the Hot Seat: Leaders and Teams for Critical Incident Management*. Chichester, UK: Wiley.

Gailbraith, J. 1979. *Organizational Design*. Reading, MA: Addison-Wesley.

Garfinkel, H. ed. 1986. *Ethnomethodological Studies of Work*. London, UK: Routledge.

Gherardi, S. and D. Nicolini. 2002. Learning the trade: A culture of safety in practice. *Organization* 9:191–223.

Gherardi, S. 2009. *Organizational Knowledge: The Texture of Workplace Learning*. Malden, MA: John Wiley and Sons.

Goffman, E. 1960. *The Presentation of the Self in Everyday Life*. London, UK: Penguin.

Guirdham, M. 1990. *Interpersonal Skills at Work*. Englewood Cliffs, NJ: Prentice Hall.

Hager, P. 2004. Conceptions of learning and understanding learning at work. *Studies in Continuing Education* 26:3–17.

Hall, R. 2002. Enterprise resource planning systems and organizational change: Transforming work organization? *Strategic Change* 11:263–270.

Hannan, G. 1996. Selecting air traffic controllers: Airservices Australia's ATC conversion course selection program. In *Applied Aviation Psychology: Achievement, Change Challenge*, eds. B. J. Hayward and A. R. Low. Aldershot, UK: Avebury Aviation.

Harper, C. 2015. *Organizations: Structures, Processes and Outcomes*. Oxfordshire, UK: Routledge.

Hartel, C. E. and G. F. Hartel. 1995. *Controller Resource Management: What Can We Learn from Aircrews?* U.S. Department of Transportation, Federal Aviation Administration.

Hartley, J. F. 1996. Intergroup relations in organizations. In *Handbook of Work Group Psychology*, ed. M. A. West. London, UK: John Wiley and Sons.

Heath, C. and P. Luff. 1991. Collaborative activity and technological design: Task coordination in London underground control rooms. *Paper presented at the Second European Conference on Computer-supported Cooperative Work*, Amsterdam, The Netherlands.

Hedberg, B. 1981. How organizations learn and unlearn. *Cited in Organizational Learning in the Global Context*, eds. M. L. Brown, M. Kenney and M. J. Zarkin. 2006. Aldershot, UK: Ashgate Publishing, Ltd.

Helmreich, R. L. and H. C. Foushee. 1993. Why crew resource management? Empirical and theoretical bases of human factors training in aviation. In *Cockpit Resource Management*, eds. E. L. Weiner, B. G. Kanki and R. L. Helmreich. San Diego, CA: Academic Press.

Hendry, C. 1996. Understanding and creating whole organizational change through learning theory. *Human Relations* 49:621–640.

Hendry, J. 2006. *Between Enterprise and Ethics: Business and Management in a Bimoral Society*. Oxford, UK: Oxford University Press.

Hill, C. W., G. R. Jones and M. A. Schilling. 2014. *Strategic Management Theory: An Integrated Approach*. Cengage Learning.

Hodge, S. 2016. After competency-based training: Deepening critique, imagining alternatives. *International Journal of Training Research*. 14:171–179.

Holland, D., D. Lachicotte, D. Skinner and C. Cain. 1998. *Identity and Agency in Cultural Worlds*. Cambridge, UK: Harvard University Press.

Holland, D. and J. Lave. 2001. *History in Person*. Santa Fe, NM: SAR Press.

Hoc, J. M. and X. Carlier. 2002. Role of a common frame of reference in cognitive cooperation: Sharing tasks between agents in air traffic control. *Cognition, Technology and Work* 4:37–47.

Hollnagel, E. and D. D. Woods. 2005. *Joint Cognitive Systems: Foundations of Cognitive Systems Engineering*. Boca Raton, FL: CRC Press.

Hughes, J. A., D. Randall and D. Shapiro. 1992. Faltering from ethnography to design. *Paper presented at the ACM 1992 conference on Computer-supported Cooperative Work*, Toronto, Canada.

Hutchins, E. and T. Klausen. 1996. Distributed cognition in an airline cockpit. In *Cognition and Communication at Work*, eds. Y. Engestrom and D. Middleton. New York, NY: Cambridge University Press.

Issac, A. R. and B. Ruitenberg. 1999. *Air Traffic Control: Human Performance Factors*. Aldershot, UK: Ashgate Aviation.

James, P. 1997. Models of vocational development revisited: Reflecting on concerns. In *Teaching and Learning in Vocational Education and Training*, ed. R. Blunden. Katoomba, Australia: Social Science Press.

Joas, H. 1996. *The Creativity of Action*. Cambridge, UK: Polity Press.

Jones, G. 1995. *Organizational Theory: Text and Cases*. Reading, MA: Addison-Wesley.

Kaptelinin, V. and B. A. Nardi. 2006. *Acting with Technology: Activity Theory and Interaction Design*. Boca Raton, FL: MIT press.

Kimme, M. 2008. *Guyland: The Perilous World Where Boys Become Men*. New York, NY: Harper.

Klein, R. L., G. A. Bigley and K. H. Roberts. 1995. Organizational culture in high reliability organizations: An extension. *Human Relations* 48:771–793.

Knight, K. H., M. H. Elfenbein and M. B. Martin. 1997. Relationship of connected and separate knowing to the learning styles of Kolb: Formal reasoning and intelligence. *Sex-Roles: A Journal of Research* 375:401–414.

Kolb, D. 1984. *Experiential Learning*. Englewood Cliffs, NJ: Prentice Hall.

Kornbluh, H. and R. Greene. 1989. Learning, empowerment and participative work processes: The educative work environment. In *Socialization and Learning at Work: A New Approach to the Learning Process in the Workplace and Society*, eds. H. Leymann and H. Kornblu. Brookfield, VT: Gower.

Kotter, J. P. 2008. *Corporate Culture and Performance*. New York, NY: Simon and Schuster.

Latour, B. 1987. *Science in Action*. Cambridge, MA: Harvard University Press.

Lave, J. 1996. The practice of learning. In *Understanding Practice: Perspectives on Activity and Context*, eds. S. Chaiklin and J. Lave. New York, NY: Cambridge University Press.

Lave, J. and E. Wenger. 2005. Practice, person, social world. In *An Introduction to Vygotsky 2*, ed. H. Daniels, 149–156. New York, NY: Routledge.

Lawrence, B. S. 2006. Organizational reference groups: A missing perspective on social context. *Organization Science* 17:80–100.

Lawrence, P. R. and J. W. Lorsch. 1969. *Organization and Environment: Managing Differentiation and Integration*. Homewood, IL: Irwin.

Lewin, K. 1951. *Field Theory in Social Science*. New York, NY: Harper Collins.

Lois, J. 2001. Peaks and valleys: The gendered emotional culture of edgework. *Gender and Society* 15:381–406.

Loui, M. R. 1986. An involvestigator's guide to workplace culture. *Perspectives* 3:73–93.

Macphee, I. 1992. *Independent Review of the Civil Aviation Authority's Tender Evaluation Process for the Australian Advanced Air Traffic System*. Canberra, Australia: Australian Government Publishing Service.

Malakis, S., T. Kontogiannis and B. Kirwan. 2010. Managing emergencies and abnormal situations in air traffic control part II: Teamwork strategies. *Applied Ergonomics* 41:628–635.

Mason, J. 1993. Learning from experience in mathematics. In *Using Experience for Learning*, eds. D. Boud, R. Cohen and D. Walker. Buckingham, UK: Open University Press.

Maurino, D. E., J. Reason, N. Johnston and R. B. Lee. 1995. *Beyond Aviation Human Factors: Safety in High Technology Systems*. Aldershot, UK: Avebury Aviation.

Moshansky, V. P. 1992. Commission of inquiry into the air Ontario accident at Dryden, Ontario: Final report. In *Beyond Aviation Human Factors: Safety in High Technology Systems*, eds. D. E. Maurino, J. Reason, N. Johnston and R. B. Lee. Aldershot, UK: Avebury Aviation.

Murphy, M. 1980. Review of aircraft incidents. In *Resource Management on the Flight Deck: Proceedings of a NASA/Industry Workshop*, eds. G. Cooper, M. D. White and J. K. Lauber. Moffet Field, CA: NASA–AMES Research Centre.

Nonaka, I. and H. Takeuchi. 1995. *The Knowledge-Creating Company: How Japanese Companies Create the Dynamics of Innovation*. New York, NY: Oxford University Press.

Orr, J. E. 1996. *Talking About Machines: An Ethnography of a Modern Job.* Ithaca, NY: Cornell University Press.

Owen, C. and W. Page. 2010. The reciprocal development of expertise in air traffic control. *International Journal of Applied Aviation Studies* 10:131.

Owen, C. 2013. Gendered communication and public safety: Women, men and incident management. *Australian Journal of Emergency Management* 28(2):4–15.

Oxford Dictionary. 2014. *The Concise Oxford Dictionary of Current English.* Oxford, UK: Clarendon Press.

Patton, M. 1990. *Evaluation and Research Methods.* Newbury Park, CA: Sage.

Pea, R. D. 1993. Practices of distributed intelligence and designs for education. In *Distributed Cognitions: Psychological and Educational Processes,* ed. G. Salomon. Cambridge, MA: Cambridge University Press.

Piaget, J. 1960. *The Child's Conception of the World.* Totowa, NJ: Littlefield.

Powell, W. W. and K. Snellman. 2004. The knowledge economy. *Annual Review of Sociology* 30:199–220.

Putnam, R. T. and H. Borko. 1997. Teacher learning: Implications of new views of cognition. In *International Handbook of Teachers and Teaching,* ed. B. J. Biddle. Dordrecht, The Netherlands: Kluwer Academic Press.

Rantanen, E. M., S. J. Yeakel and K. S. Steelman. 2006. En route controller task prioritization research to support CE-6 human performance modeling: Analysis of high-fidelity simulation data, Phase II. University of Illinois.

Ratner, R. 1992. Report of the 1992 review of the Australian Air Traffic Services System. *Department of Transport and Communications and Civil Aviation Authority of Australia.* Canberra, Australia: Bureau of Air Safety Investigation.

Reason, J. 1998. *Managing the Risks of Organizational Accidents.* Aldershot, UK: Ashgate.

Redding, R. E., J. R. Cannon and T. L. Seamster. 1992. Expertise in air traffic control (ATC): What is it, and how can we train for it? In *Proceedings of the Human Factors and Ergonomics Society Annual Meeting* 36(17):1326–1330. Los Angeles, CA: SAGE Publications.

Resnick, L. 1993. Shared cognition: Thinking as social practice. In *Perspectives on Socially Shared Cognition,* eds. L. B. Resnick, J. M. Levine and S. D. Teasley. Washington DC: American Psychological Association.

Resnick, L. B., J. M. Levine and S. D. Teasley, eds. 1993. *Perspectives on Socially Shared Cognition.* Washington: American Psychological Association.

Rogoff, B. 1990. *Apprenticeship in Thinking: Cognitive Development in Social Context.* New York, NY: Oxford University Press.

Ryan, G. W., H. R. Bernard, N. Denzin and Y. Lincoln. 2000. *Handbook of Qualitative Research.* Thousand Oaks, CA: Sage.

Salomon, G. ed. 1993. *Distributed Cognitions.* New York, NY: Cambridge University Press.

Sambrook, S. 2005. Factors influencing the context and process of work-related learning: Synthesizing findings from two research projects. *Human Resources Development International* 8:101–119.

Sawchuk, P. 2008. Theories and methods for research on informal learning and work: Towards cross-fertilization. *Studies in Continuing Education* 30:1–16.

Scheeres, H. and C. Rhodes. 2006. Between cultures: Values, training and identity in a manufacturing firm. *Journal of Organizational Change Management* 19:223–237.

Schein, E. H. 1996. Culture: The missing concept in organization studies. *Administrative Science Quarterly* 41:229–240.

Schein, E. H. 2009. *The Corporate Culture Survival Guide.* Hoboken, NJ: John Wiley and Sons.

Schon, D. A. 1983. *The Reflective Practitioner: How Professionals Think in Action.* New York, NY: Basic Books.

Schon, D. A. 1987. *Educating the Reflective Practitioner.* San Francisco, CA: Jossey-Bass.

Schon, D. A. ed. 1991. *The Reflective Turn.* New York, NY: Teachers College Press.

Schultz, K. P. 2005. Learning in complex organizations as practicing and reflecting: A model development and application from a theory of practice perspective. *Journal of Workplace Learning* 17:493–507.

Shaiken, H. 1996. Experience and the collective nature of skill. In *Cognition and Communication at Work*, eds. Y. Engestrom and D. Middleton. New York, NY: Cambridge University Press.

Sharma, M. and A. Ghosh. 2007. Does team size matter? A study of the impact of team size on the transactive memory system and performance of IT sector teams. *South Asian Journal of Management* 14:96.

Sheen, M. J. 1987. MV Herald of Free Enterprise. *Report of Court* 8074. London, UK: Department of Transport.

Shulman, L. S. 2005. Signature pedagogies in the professions. *Daedalus* 134:52–59.

Smiricich, L. 1983. Concepts of culture and organizational analysis. *Administrative Science Quarterly* 28:339–358.

Soraji, Y., K. Furuta, T. Kanno et al. 2012. Cognitive model of team cooperation in en-route air traffic control. *Cognition, Technology and Work* 14:93–105.

Strauss, A. and Corbin, J. 1998. *Basics of Qualitative Research: Procedures and Techniques for Developing Grounded Theory.* 2nd Edition. London, UK: Sage.

Suchman, L. 1987. *Plans and Situated Actions: The Problem of Human-Machine Communication.* Cambridge, UK: Cambridge University Press.

Suchman, L. 1996. Constituting shared workspaces. In *Cognition and Communication at Work*, eds. Y. Engestrom and D. Middleton. New York, NY: Cambridge University Press.

Taylor, J. L., R. O'hara, M. S. Mumenthaler, A. C. Rosen and J. A. Yesavage. 2005. Cognitive ability, expertise, and age differences in following air-traffic control instructions. *Psychology and Aging* 20(1):117–133.

Tesch, R. 1990. *Qualitative Research: Analysis Types and Software Tools.* New York, NY: Falmer Press.

Timma, H. 2007. Experiencing the workplace: Shaping workers identities through assessment, work and learning. *Studies in Continuing Education* 29:163–179.

Tosi, H. L. and M. Pilati. 2011. *Managing Organizational Behaviour: Individuals, Teams and Organizations.* Cheltenham, UK: Edward Elgar.

Trice, H M. and J. M. Beyer. 1984. Studying organization culture through rites and ceremonials. *Academy of Management Review* 9:653–669.

Tsang, E. W. and S. A. Zahra. 2008. Organizational unlearning. *Human Relations* 61:1435–1462.

Van de Ven, A. H., M. Ganco and C. R. Hinings. 2013. Returning to the frontier of contingency theory of organizational and institutional designs. *The Academy of Management Annals* 7:393–440.

Vaughan, D. 1997. The trickle-down effect: Policy, decisions, risky work and the Challenger tragedy. *California Management Review* 39:80–102.

Vaughan, D. 2006. NASA Revisited: Theory, analogy and public sociology 1. *American Journal of Sociology* 112:353–393.

Von Glinow, M. and S. Mohrman, eds. 1990. *Managing Complexity in High Technology Organizations*. New York, NY: Oxford University Press.

Vygotsky, L. S. 1978. *Mind in Society*. Cambridge, MA: Harvard University Press.

Watkins, K. E. and V. J. Marsick. 1993. *Sculpting the Learning Organization: Lessons in the Art and Science of Systemic Change*. San Francisco, CA: Jossey-Bass.

Weick, K. E. 1987. Organizational culture as a source of high reliability. *California Management Review* 29:112–127.

Weick, K. E. 2012. *Making Sense of the Organization, Volume 2: The Impermanent Organization*. Chichester, UK: John Wiley and Sons.

Wellins, R. S., W. C. Byham and J. M. Wilson. 1991. *Empowered Teams: Creating Self-Directed Work Groups that Improve Quality, Productivity and Participation*. San Fransciso, CA: Jossey-Bass.

Wenger, E., R. A. McDermott and W. Snyder. 2002. *Cultivating Communities of Practice: A Guide to Managing Knowledge*. Boston, MA: Harvard Business Press.

Wertsch, J. 1998. *Mind as Action*. New York, NY: Oxford University Press.

Westrum, R. 1993. Cultures with requisite imagination. In *Verification and Validation of Complex Systems: Human Factors Issues*, eds. J. A. H. Wise, V. D. Hopkin and P. Stager. 401–416. Berlin, Germany: Springer-Verlag.

Westrum, R. 1997. Social factors in safety critical systems. In *Human Factors in Safety-Critical Systems*, eds. F. Redmill and J. Rajan. Oxfordshire, UK: Butterworth–Heineman.

Westrum, R. 1998. Organizational learning in the aviation business. Paper presented at the *Fourth Aviation Psychology Symposium*, Manly.

Wheelan, S. A. 1994. *Group Processes: A Developmental Perspective*. Boston, MA: Allyn and Bacon.

Wolfe, T. 1979. *The Right Stuff*. London, UK: Picador.

Zuboff, S. 1988. *In the Age of the Smart Machine: The Future of Work and Power*. New York, NY: Basic Books.

Index